HOUSE DUST MITES

HOUSE DUST MITES

Natural History, Control, and Research Techniques

ROB DE BOER

Academic Press is an imprint of Elsevier
125 London Wall, London EC2Y 5AS, United Kingdom
525 B Street, Suite 1650, San Diego, CA 92101, United States
50 Hampshire Street, 5th Floor, Cambridge, MA 02139, United States
The Boulevard, Langford Lane, Kidlington, Oxford OX5 1GB, United Kingdom

ISBN 978-0-443-19111-4

For information on all Academic Press publications
visit our website at https://www.elsevier.com/books-and-journals

Publisher: Nikki P. Levy
Acquisitions Editor: Kelsey Connors
Editorial Project Manager: Anthony Marvullo
Production Project Manager: Omer Mukthar
Cover Design: Fanny de Boer and Vicky Pearson Esser

Typeset by STRAIVE, India

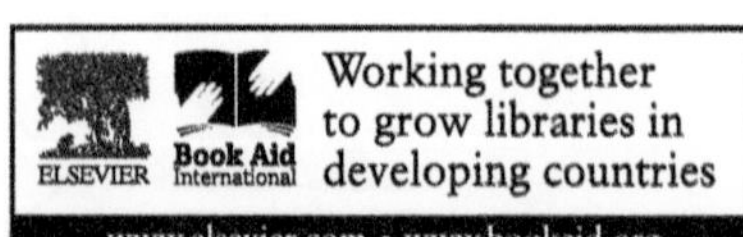

Dedication

For Winnie, Cuno, and Fanny

Contents

Preface

During the years 1975 until 2000, I was, with interruptions, employed by the University of Amsterdam. My workstation was the Laboratory of Experimental Entomology. The laboratory has been renamed several times, but the word "acarology" never formed part of the name. Yet most of the research going on there concerned mites. I worked on spider mites, ticks, and, during the years 1987 through 2000, on house dust mites. For my employment I depended on financial support from external funding organizations. Satisfying the demands of these organizations had priority of course. There was little time to go into interesting sideways or delve into underlying, more theoretical questions. Eventually lack of funds forced me to stop working on dust mites. But the fascination was still there. Through the scientific literature I had meticulously followed the ongoing research on dust mites and I felt the need to organize and summarize the overwhelming accumulation of information and to clarify topics that are not so easy to understand. Not before my retirement in 2012, I found time for an attempt to do that. It was only an attempt but, I believe, not a complete failure. In fact I was quite content with it myself.

Explaining something to someone else is often the best way to test whether you really understand it yourself. I had nobody. But an anonymous reader is a rather satisfactory alternative. So I wrote this book primarily for myself. But who else could benefit?

Dust mites are important for human health, albeit in a negative way. That is why they received much attention of acarologists. It is quite rare that so much research has been done and so much information has accumulated concerning one single animal species. Mites are tiny organisms, living without glamour, but living. Mites in general never received much attention of the general public. In fact they are quite obscure creatures for most people. Yet they have some remarkable, even unique, properties interesting for anyone with an interest in natural history. I tried to write a text that is understandable for everybody: outsiders as well as insiders. Medical professionals and the allergy patients themselves may wish to know some more about these secretive creatures that constitute a serious health hazard. Also undergraduate biology students who wish to learn more about mites could benefit. But above all, this is what I had in mind as a reader: an acarologist, either familiar or not familiar with the subject, who I try to convince of the need to perform certain investigations on house dust mites.

Acknowledgments

The Netherlands Asthma Foundation (NAF), now called "Het Longfonds," supported my research financially during the years 1980–90, allowing me to extend my knowledge on house dust mite biology. All the time I highly appreciated the help, both technical and intellectual, of my friend Kees Kuller. During the years when I was writing, my wife Winnie bore the brunt of my absentmindedness, my son, Cuno, again and again helped me out when I was stuck in a software predicament and my daughter, Fanny, made invaluable contributions to many of the illustrations of the book. Scans of illustrations originating from published articles were always readily provided by staff members of the Naturalis library in Leiden.

I alone, of course, am responsible for errors, omissions, and other shortcomings.

CHAPTER ONE

An introductory note on the medical significance of dust mites

1.1 Discoveries

It has often been said that Frits Spieksma discovered the house dust mite. That is not exactly true. How could it be? Formally the European house dust mite is called *Dermatophagoides pteronyssinus* (Trouessart, 1897). That means that this mite was described as a distinct species in 1897 by an acarologist called Trouessart before Spieksma was born (in 1936). It had been known for a long time that asthma attacks can be triggered in susceptible people when they inhale air that is somehow contaminated with house dust, for instance during house cleaning activities. "Sweeping causes choking" (Kern, 1921). The Flemish physician John Baptista Van Helmont (1662) may have been the first to mention the connection between asthma attacks and exposure to dust. He described the case of a monk who fell short of breath any time a place was swept or dust was stirred up by any other means.

Skin tests (Fig. 1.1) were done at least since 1917 (Walker, 1917). Asthmatic people show a positive skin reaction to extracts of house dust far more often than nonasthmatics. Some constituent of house dust acts as an allergen in these people. In 1928, Dekker (1928), a German physician, published an article in which he suggested that mites could be the source of this allergen. Skin tests can be deployed to determine the presence of an allergen in an extract in a semiquantitative way. A higher concentration of the allergen will result in a stronger skin reaction as can be judged by the size and appearance of the weal (Fig. 1.1). Spieksma, working in Leiden, The Netherlands, set up cultures of *Dermatophagoides pteronyssinus* and prepared extracts of these mites. Skin tests on susceptible and nonsusceptible individuals using these extracts besides house dust extracts from various sources strongly pointed to the mites as the source of the allergen. The experiments were described

House Dust Mites
https://doi.org/10.1016/B978-0-443-19111-4.00017-4

in Spieksma's dissertation (1967) and also in a book by Voorhorst et al. (1969). The book was epoch making. It spurred a host of medical studies as well as biological research on this mite—a mite species that was hardly noticed before.

I suspect that today among medical professionals the prevailing notion is that, by now, we must know everything there is to be known about these simple creatures. The truth is that we do not fully understand why some houses have a thriving mite population, whereas quite similar dwellings in the same neighborhood are almost devoid of mites and mite allergens.

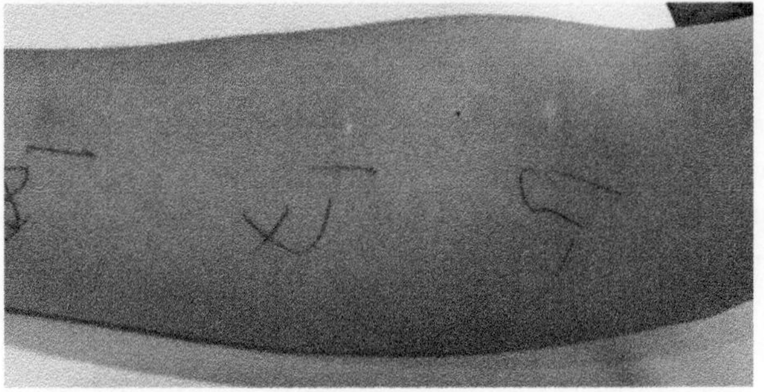

Fig. 1.1 Skin prick test. *(Photo: Shutterstock.com.)*

Table 1.1 Dust mite allergens and their properties.

Allergen	Frequency of reactivity (%)	Molecular mass (kDa)	Number of amino acids	Function
Der p 1	>90	25	222	Cysteine protease
Der p 2	>90	14	129	Not known
Der p 3	51->90	25	?	Trypsin
Der p 4	25–46	56	?	Amylase
Der p 5	>55	13	113	Not known
Der p 6	39	30	?	Chymotrypsin
Der p 7	53	22	198	Not known
Der p 8	?	26	219	Not known
Der p 9	>90	28	?	Serine protease
Der f 1	79	25	223	Cysteine protease
Der f 2	83	14	129	Not known
Der f 3	42–70	30	232	Trypsin
Der f 6	31	30	?	Chymotrypsin
Der f 7	46	22	196	Not known
Der f 10	81	33	284	Tropomyosin

From Stewart, G.A., Thompson, P.J., 1996. The biochemistry of common aeroallergens. Clin. Exp. Allergy 26, 1020–1044.

1.2 Identification and quantification

Dust mites turned out to be abundant in almost every human dwelling. Spieksma (1967) found it "astonishing" that this had always escaped the attention of investigators. Soon after the publication of the book by Voorhorst, Spieksma, and Varekamp a great number of epidemiological studies were initiated. These studies focused on the link between allergy symptoms and the abundance of mites in houses. Among the earliest studies were those by Jens Korsgaard, a lung physician working in Aarhus, Denmark. He found that the number of mites in dust from the houses of 25 patients with newly diagnosed asthma was much greater than in the homes of 75 randomly selected matched control subjects (Korsgaard, 1983). Besides *D. pteronyssinus* two or three other species of mites were very common in house dust from all over the world. In particular *Dermatophagoides farinae*, the so-called American house dust mite, was quite common and, despite its name, not only in America. Korsgaard took the number of mites per gram of vacuumed dust as a measure of exposure to mite allergens. The allergens themselves were not yet identified. But that was only a matter of time. Later investigators often took the concentration of Der p 1 + Der f 1 per gram of dust as a measure of exposure. Der p 1 and Der f 1 are proteins in the feces of *D. pteronyssinus* and *D. farinae*, respectively, and regarded as major house dust mite allergens (Table 1.1). These proteins are digestion enzymes. Actually it is surprising that these proteins emerged as inhalation allergens because the fecal pellets of dust mites are relatively large, much larger than, for instance, the particles carrying the allergens of cats and dogs. Vigorous disturbance of dust collectors like carpets, rugs, stuffed furniture, and mattresses is necessary to render the mite feces airborne in the first place and then the fecal pellets will rather quickly fall back to the ground. This was confirmed by studies in which airborne sampling was done (for instance, Swanson and Agarwa, 1985; Swanson et al., 1989, Custovic et al., 1995). Thus the exposure to dust mite allergens probably occurs erratically, in brief episodes with high concentrations in the inhaled air.

From a medical point of view, airborne sampling provides the most relevant measure of allergen exposure, i.e. the amount of allergen that is inhaled by the inhabitants of a house during the day and during the night. But sampling air for an extended period is tedious. A more practical sampling method would be highly desirable. Therefore the amount of allergen per gram of vacuumed dust has been chosen as the preferred measure of allergen exposure. There must be a correlation between the concentration of allergen in house dust and the amount of allergen in the air following a disturbance of settled dust. Of course this is a compromise between the most desirable and the most practical of sampling procedures. Imagine, for instance, that the inhabitants of a house come home from a visit to the beach and bring in sand on their clothes which mixes with the floor dust. The concentration of allergen per gram of vacuumed material will then be drastically changed by the presence of the heavy sand grains. But these grains will not have any effect on the amount of allergen in the air. Nevertheless, it is assumed that, in general, the concentration of allergen in settled dust provides a reasonably good indication of the burden of allergen exposure.

Mite feces also contain a lot of guanine, a metabolic waste product. As expected, there is a strong correlation between the concentration of guanine and the concentration of mite allergens. A semiquantitative test for guanine content of house dust, the "Acarex test," can be used, even by lay people, to obtain an indication of the allergen burden in their homes (see Fig. 1.3).

1.3 Exposure to mite allergens and disease

The abovementioned study by Korsgaard already showed a connection between the number of mites in house dust and allergy symptoms. As soon as immunoassays for the detection and quantification mite allergens had been developed, investigators no longer determined the number of mites. Instead the concentration of the mite allergens Der p 1 or the sum of Der p 1 + Der f 1 was taken as a measure of allergen exposure. Examples are the study by Sporik et al. (1990) in southern England and the study by Vervloet et al. (1991) in southern France. Vervloet and his cooperators studied 49 asthmatic subjects attending an outpatient clinic in Marseille in 1987/1988. These were subdivided into three groups on the basis of their medication requirements, which of course reflects the severity of their symptoms. It was found that patients with the severest symptoms generally had the highest concentration of house dust mite allergens in the dust from their mattresses (Fig. 1.4). Sporik et al. (1990) studied some 70 newborn children of which one parent had either asthma or hay fever and therefore had an increased chance of becoming atopic also. The children were followed until age 11. One of the findings was that asthma symptoms started at an earlier age when the child lived in a home where allergen levels were high ($P < .015$).

Also, in a tropical country like Zambia, asthma is very common and asthmatics were found to be sensitized for house dust mite allergens much more often than healthy control subjects (Buchanan and Jones, 1972).

In the temperate climatic regions, dust mites and their allergens rise to a peak in the autumn and then fall again during the winter. And indeed an aggravation of respiratory symptoms is observed during the autumn (Heide et al., 1997). Circumstantial evidence indicates that this worsening of symptoms is directly caused by the increased exposure to allergens and not to some other environmental change that is associated with the season.

Findings such as those outlined before suggest that it is desirable and useful to take measures aiming to avoid exposure to mite allergens. Zero exposure is not realistic. All places where people go are contaminated to some degree. Allergens are transported on the clothes of people and spread all over. Even brand new carpets and mattresses have been found to contain measurable quantities of mite allergen (van der Hoeven et al., 1992; de Boer, 2002). The question is: What is the highest concentration of allergens in house dust that can be regarded as harmless? Or can even the lowest concentration do some harm? In order to address this and other questions, several international expert meetings were organized during the years 1970–80. At one of these meetings (Platts-Mills et al., 1991, 1992) the consensus was reached that:

(1) the threshold level for sensitization is:

$2\,\mu g$ Der p 1 + Der f 1 per gram of dust or 100 mites per gram of dust. and

(2) the threshold level for acute asthma attacks is:

$10\,\mu g$ Der p 1 + Der f 1 per gram of dust or 500 mites per gram of dust. However, indications have been found that even levels below $2\,\mu g/g$ of dust can be sufficient to induce sensitization (Peat et al., 1993; Munir et al., 1993).

1.4 BHR, FEV, and PC-20

Inhaling an allergen seems to have two effects: (1) It may cause an acute attack of asthma. (2) It may cause an increase of 'bronchial hyperreactivity' (BHR). Hyperreactive lungs react to various stimuli, not only to specific allergens but also to other irritating substances in the inhaled air. The degree of BHR can be quantified as follows: First the "Forced Expiratory Volume" (FEV) of the lungs of the patient is determined using a spirometer (Fig. 1.2). Then an aerosol with a low concentration of a nonspecific irritant (e.g., histamine or methacholine) is inhaled and the FEV is determined again. This procedure is repeated with an increased dose of the irritant and repeated again and again until a 20% fall of the FEV is attained. The concentration of the irritant that is needed to cause a fall of 20% is called the PC-20 and is regarded as a measure of BHR. For hyperreactive lungs only a very little of the irritant is sufficient to make the patient wheeze. Thus a low PC-20 is bad, a high one is good.

1.5 Secondary prevention: Yes, primary prevention: No

Ehnert et al. (1992) did an intervention experiment. Twenty-four asthmatic children were subdivided into three groups. In one group the mattresses and the bedding were encased in allergen–impermeable covers and the carpets were treated with a substance (tannic acid) that denatures the allergens. The beds of the second group received a different treatment to reduce allergen levels. Here a foam containing the mite killing agent benzyl benzoate was applied. The beds of the third group were treated with a foam containing no benzyl benzoate, i.e., a placebo. This group served as a blind control. The results were presented graphically (Fig. 1.5) and the differences in pattern were quite striking. Only in the group with the encased mattresses the allergen concentrations in dust samples were strongly reduced. And only in this group an amelioration of BHR was seen. The observed changes were especially marked in the individuals with severest BHR. The number of patients with PC-20 < 0.5 mg histamine/ml declined from 5 (at the start) to four (at 4 months) to 1 (at 8 months) and finally to 0 (at 12 months). In the placebo group and in the group where the acaricidal foam was applied (with little effect on allergen concentrations) these numbers remained the same all along (4 and 6, respectively). The numbers of patients involved in this study were not very big and the statistical tests applied by the authors revealed only modest statistical significance. But the observations are at least encouraging, offering hope that health improvement can be achieved through allergen control measures.

I emphasize that this study is just one example and I could be accused of data snooping and cherry-picking. Many other studies with a similar objective have been done, but not always with a similarly convincing result. A problem with these intervention studies is that it is impossible to completely eliminate all the allergen from the patients' environment. If no effect can be shown, it may be that the allergen reduction was simply not strong enough. Colloff (2009) in his book entitled *Dust Mites*

(583 pp.) published in 2009 scrutinized a large number of clinical trials. He identified 30 studies that met his inclusion criteria (randomized controlled trials of allergen avoidance intervention in homes of patients with allergic asthma, with parallel control and intervention groups and with quantitative measurement of allergen and/or mites). In only 12 of these 30 trials, the active group showed a significantly greater health improvement than the control group. In accordance with these positive clinical findings, the allergen levels in the active group, but not in the control group, were indeed strongly reduced. And that was not so in the 18 trials with negative outcomes. Here the allergen reduction in the active group was only small. Colloff concludes: "These data suggest that no clinical improvement in asthma occurs without a reduction in allergen levels of at least two-fold."

Trying to prevent allergy symptoms by avoiding contact with a specific allergen is a strategy that must be classified as "secondary prevention" or, in a more refined subdivision, "tertiary prevention," because the patient is already atopic and sensitized for that allergen. Of course it would be much better if it could be prevented that a child develops an atopic condition in the first place. That would be classified as "primary prevention." Unfortunately, attempts to demonstrate the possibility to prevent the development of atopy through allergen avoidance have met with disappointing results. It seems that other factors than exposure to allergens determine whether a person becomes an allergy patient or not.

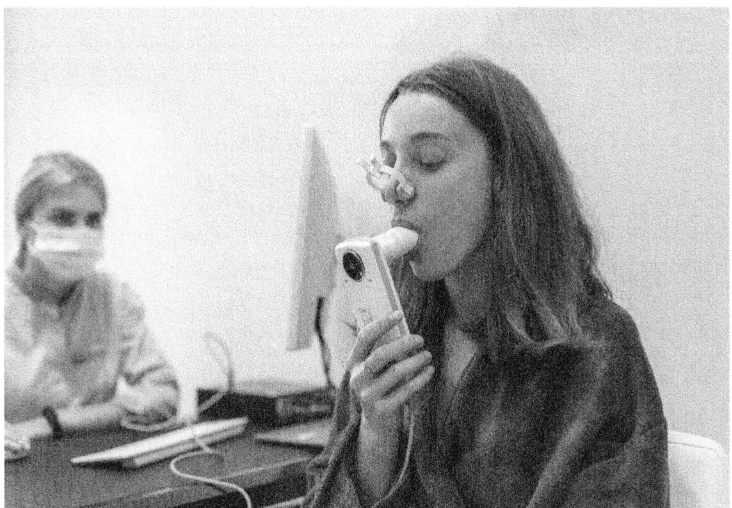

Fig. 1.2 The spirometer. *(Photo: Shutterstock.com.)*

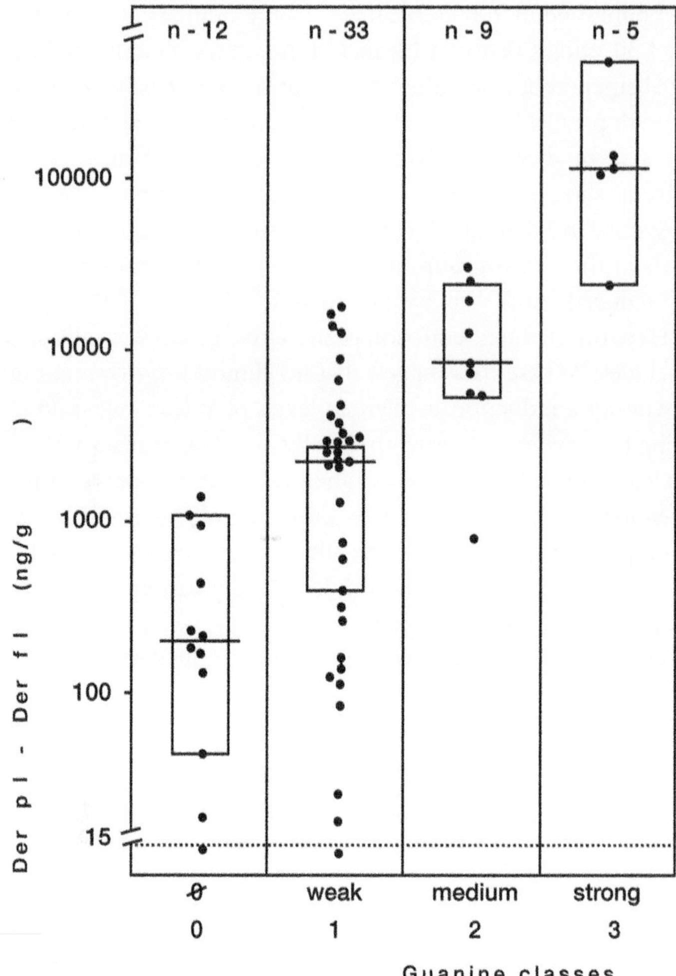

Fig. 1.3 The correlation between the allergen content of settled dust and the concentration of guanine as determined by the "Acarex test." *(From Lau-Schadendorf Rusche, A., Weber, A. Werthmann, I., Büttner-Götz, P., Wahn, U., 1990. Nachweis van Hausstaubmilbenallergenen-ELISA und Guaninbestimmung im Vergleich. Allergologie 13: 12–15.)*

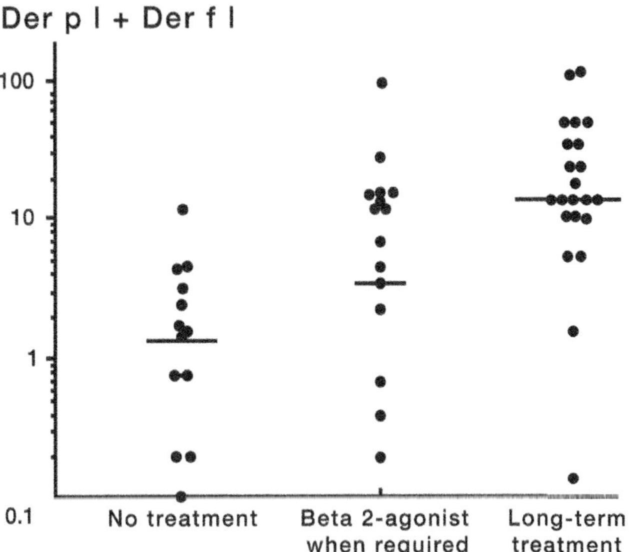

Fig. 1.4 Distribution of individual Ag P1 Eq levels (defined as the sum of major allergens Der p 1 and Der f 1) in house dust from mattresses according to the treatment required. Each patient is represented by a single dot. The horizontal bars represent the mean levels computed in each group. *(From Vervloet, D., Charpin, D., Haddi, E., N'guyen, A., Birnbaum, J., Soler, M., van der Brempt, X., 1991. Medication requirements and house dust mite exposure in mite-sensitive asthmatics. Allergy 46, 554–558.)*

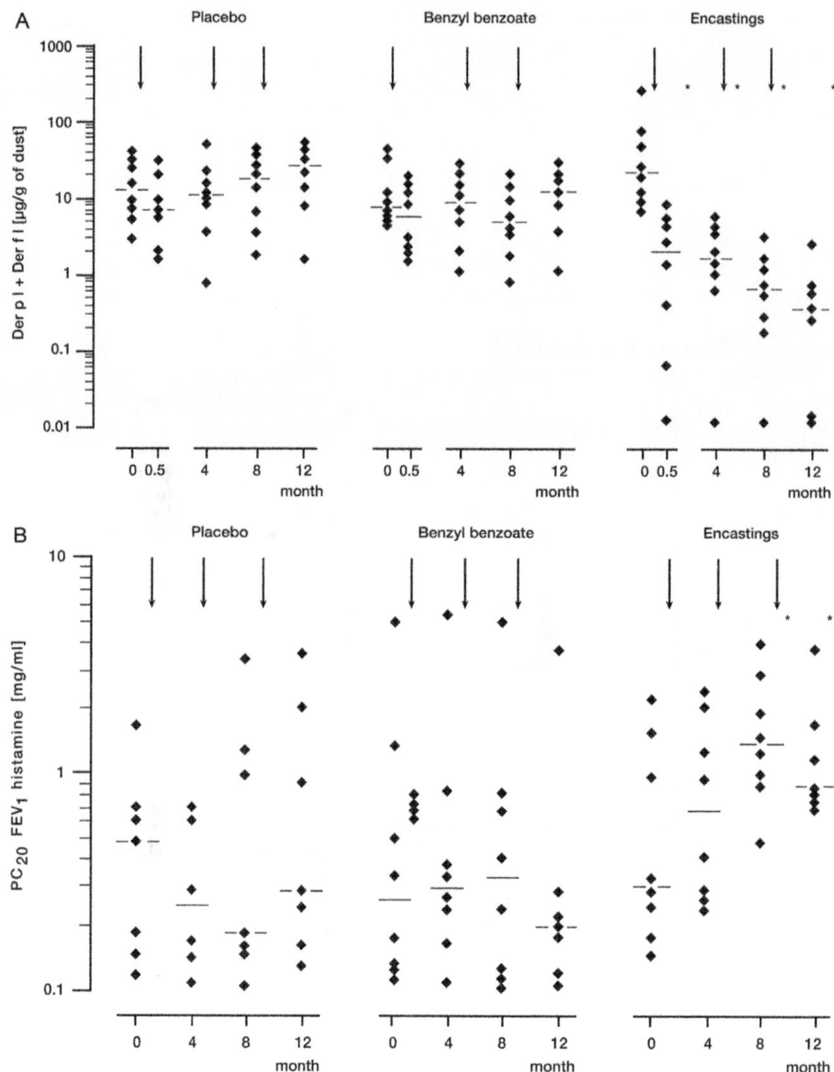

Fig. 1.5 (A) Mite allergen levels in three groups before and after treatment (arrows); *P < .05; bars: median values (Ehnert et al., 1992). (B) BHR (PC_{20} of histamine) in three groups before and after treatment (arrows); *P < .05; bars: median values. *(From Ehnert, B., Lau-Schadendorf, S., Weber, A., Buettner, P., Schou, C., Wahn, U., 1992. Reducing domestic exposure to dust mite allergen reduces bronchial hyperreactivity in sensitive children with asthma. J. Allergy Clin. Immunol. 90: 135–138.)*

References

Buchanan, D.J., Jones, I.G., 1972. Allergy to house dust mites in the tropics. Br. Med. J. 21, 764.

Colloff, M.D., 2009. Dust Mites. CSIRO/Springer Science. 583 pp.; ISBN 978-90-481-2223-3.

Custovic, A., Taggart, S.C.O., Niven, R.M., Woodcock, A., 1995. Evluating exposure to mite allergens. J. Allergy Clin. Immunol. 96, 134–135.

de Boer, R., 2002. Allergens, Der p 1, Der f 1, Fel d 1 and can f 1, in newly bought mattresses for infants. Clin. Exp. Allergy 32, 1602–1605.

Dekker, H., 1928. Asthma und Milben (reprinted in English, 1971, J. Allergy Clin. Immunol., 48, 251–252). Münchener Medizinischer Wochenschrift 75, 515–516.

Ehnert, B., Lau-Schadendorf, S., Weber, A., Buettner, P., Schou, C., Wahn, U., 1992. Reducing domestic exposure to dust mite allergen reduces bronchial hyperreactivity in sensitive children with asthma. J. Allergy Clin. Immunol. 90, 135–138.

Heide, S., van der Monchy, J.G.R., de Vries, K., de Dubois, E.J., Kauffman, H.F., 1997. Seasonal differences in airway hyperresponsiveness in asthmatic patients: relationship with allergen exposure and sensitization to house dust mites. Clin. Exp. Allergy 27, 627–633.

Kern, R., 1921. Dust sensitization in bronchial asthma. Med. Clin. North Am. 5, 751–758.

Korsgaard, J., 1983. Mite asthma and residency. Am. Rev. Resp. Dis. 128, 231–235.

Munir, A.K.M., Einarsson, R., Kjellman, N.-I.M., Björkstén, B., 1993. Mite (Der p I, Der f I) and cat (Fel d I) allergens in the homes of babies with a family history of allergy. Allergy 48, 158–163.

Peat, J.K., Tovey, E., Mellis, C.M., Leeder, S.R., Woolcock, A.J., 1993. Importance of house dust mite and Alternaria allergens in childhood asthma: an epidemiological study in two climatic regions of Australia. Clin. Exp. Allergy 23, 812–820.

Platts-Mills, T.A.E., Thomas, W.R., Aalberse, R.C., Vervloet, D., Chapman, M.D., 1991. Dust Mite allergen and asthma: Report of the second international workshop. The UCB Institute of Allergy, pp. 11–29.

Platts-Mills, T.A.E., Thomas, W.R., Aalberse, R.C., Vervloet, D., Chapman, M.D., 1992. Dust mite allergen and asthma: report of the second international workshop. J. Allergy Clin. Immunol. 89, 1046–1060.

Spieksma, F.T.M., 1967. The House-Dust Mite Dermatophagoides Pteronyssinus (Trouessart, 1897), Producer of the House-Dust Allergen. Academic thesis, Rijks Universiteit Leiden, Leiden.

Sporik, R., Holgate, S.T., Platts-Mills, T.A.E., Cogswell, J.J., 1990. Exposure to house-dust mite allergen (Der p I) and the development of asthma in childhood. N. Engl. J. Med. 323, 502–507.

Swanson, M.C., Agarwa, M.K., Reed, C.E., 1985. An immunochemical approach to indoor aeroallergen quantitation with a new volumatric air sampler: studies with mite, roach, cat, mouse, and Guinea pig antigens. J. Allergy Clin. Immunol. 76, 724–729.

Swanson, M.C., Campbell, A.R., Klauck, M.J., Reed, C.E., 1989. Correlations between levels of mite and cat allergens in settled and airborne dust. J. Allergy Clin. Immunol. 83, 776–783.

van der Hoeven, W.A.D., de Boer, R., Bruin, J., 1992. The colonisation of new houses by house dust mites (Acari: Pyroglyphidae). Exp. Appl. Acarol. 16, 75–84.

Van Helmont, J.B., 1662. Physick Refined. Lodswick Lloyd, London.

Vervloet, D., Charpin, D., Haddi, E., N'guyen, A., Birnbaum, J., Soler, M., van der Brempt, X., 1991. Medication requirements and house dust mite exposure in mite-sensitive asthmatics. Allergy 46, 554–558.

Voorhorst, R., Spieksma, F.T.M., Varekamp, H., 1969. House dust atopy and the house dust mite Dermatophagoides pteronyssinus. Stafleu's Scientific Publishing Company, Leiden, the Netherlands, 1969: #49 #34.

Walker, I.C., 1917. Studies on the sensitization of patients with bronchial asthma. J. Med. Res. 35, 497–513.

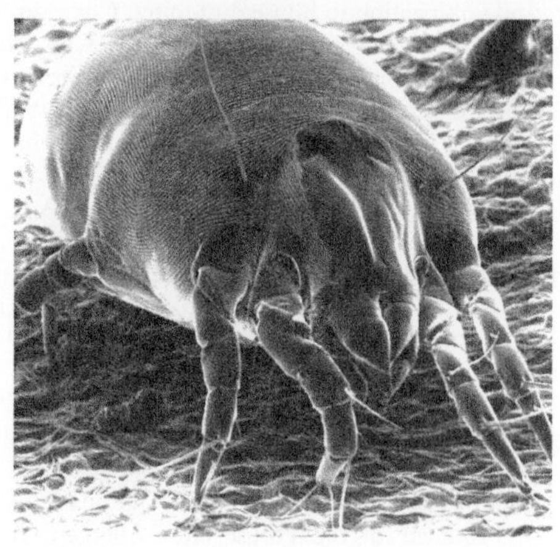

Fig. 2.1 *Dermatophagoides pteronyssinus*. SEM, adult female. *(Photo: K. Kuller.)*

CHAPTER TWO

Mites and the science of acarology

We are humans of course. No animals. But there is no doubt about it that we are mammals. Our bodies have all the diagnostic features of a mammal's body. These we share with mice, elephants, wales, bats, horses, and so on. That is how diverse the members of a single class of animals can be. Mites belong to the class *Arachnida*. These form part of yet a greater assemblage, namely the Arthropods, comparable with the Chordates to which the mammals belong. Arthropods are animals with an external skeleton and jointed legs. Many species of arthropods live in the seas: crabs, lobsters, and shrimps as well as a great diversity of pelagic forms. On the land the most frequently encountered arthropods are the insects and the spiders. Only if you look underneath stones or behind the loose bark of a decomposing, dead tree, you will see representatives of other arthropod taxa.

Spiders are a uniform group. With very few exceptions they are predators. The use of webbing is a hallmark feature. Webbing is used not only to catch prey but also to make shelters, nursery webs, egg cocoons, and draglines for walking and for ballooning. The fangs on their chelicerae are used to kill prey by a bite whereby poison is injected.

Mites are neither insects nor spiders (Figs. 2.1–2.6). Like the insects and unlike spiders, mites display a great diversity of lifestyles. Some species live on plants, either as sap suckers or as predators hunting for other mites. Others live on the feathers of birds or on and in the skin of mammals, including the skin of humans. And a vast multitude of detritus feeders live in the soils and in the nests of mammals and birds. Mites rival the insects concerning the number of species and they are equally ubiquitous. Yet, they are not so familiar to most people as the insects are. The reason must be that most species are tiny

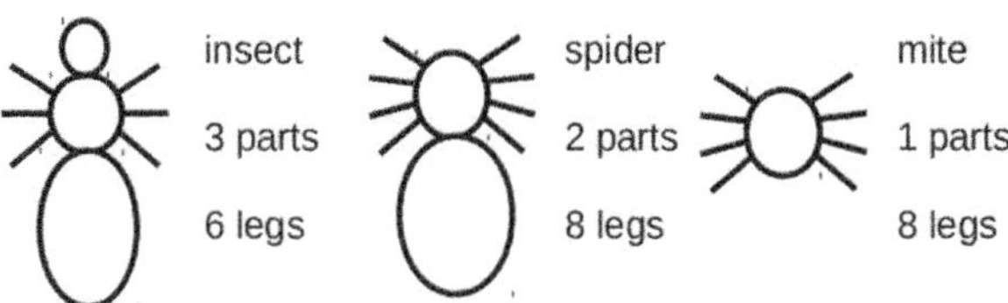

Fig. 2.2 Three dominating groups of terrestrial arthropods. *(Artwork by the author.)*

House Dust Mites
https://doi.org/10.1016/B978-0-443-19111-4.00013-7 13

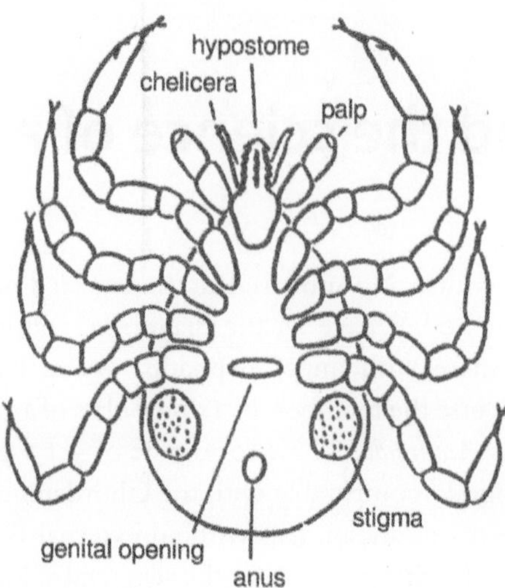

Fig. 2.3 The body plan of a tick as an example of the body plan of a mite. The large hypostome is characteristic for ticks. They use it to pierce the skin of their victims. In other mites it is much smaller. The spiracular plate has a function in breathing and is not present in house dust mites, because dust mites breathe through the skin. *(Artwork by the author.)*

creatures. Only the ticks (Fig. 2.3), a group of blood-feeding mites, are relatively large. But there is a tremendous number of very small species of mites.

Insects (Fig. 2.4) constitute a totally different group of arthropods. Some taxonomists place them not even in the same phylum with the *Acari* but assign the insects to the arthropod phylum *Uniramia*. Spiders, however, are united with the mites in the phylum *Chelicerata*, called after the chelicerae, the first pair of appendages that constitute a characteristic trait of this group (Figs. 2.5 and 2.6).

Taxonomists repeatedly change the details of their classifications in order to take novel biological findings into account. Such revisions bring along changes of the names and are often a nuisance for those who are primarily interested in the physiology or the ecology of mites and not so much in their phylogeny. The dust mite genus *Dermatophagoides* is regarded as a genus within the family *Pyroglyphidae*. Luckily, this remained so during the period after the discovery of their role in house dust allergy when a host of biological studies appeared in the literature. Also confusing synonyms never caused much trouble. Long before the dust mite investigations epoch, namely in 1864, the Russian acarologist Bogdanow described a mite and named it *Dermatophagoides scheremetewskyi*. This species is probably synonymous with *D. pteronyssinus* (Fain et al., 1990).

Fig. 2.4 *European hornet,* **Vespa crabro,** *a representative of the class Insecta.* Compound eyes and the presence of antennae are among the characters that distinguish the insects from the spiders and the mites. The compound eyes are the two crescent shaped area's on the left and the right of the head. Three simple eyes (ocelli) are also present. *(Photo: Peter Koomen (Leeuwarden).)*

Fig. 2.5 A spider (*Heteropoda tetrica*) showing the chelicerae, palps, and simple eyes. The chelicerae are the bottle shaped, frontal appendages with fangs. The slender appendages, left and right of the chelicerae, are the palps. *(Photo: Peter Koomen (Leeuwarden).)*

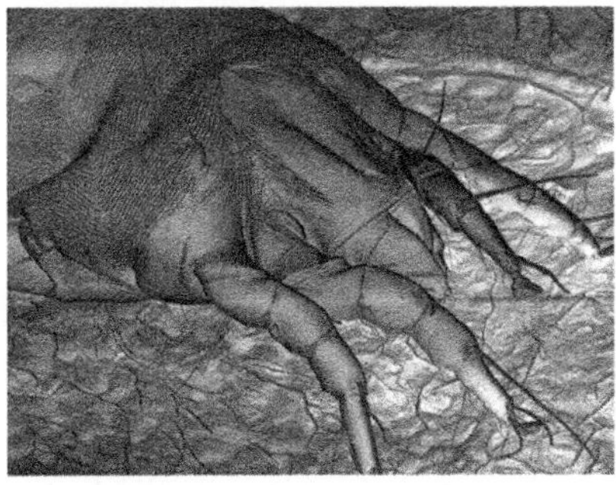

Fig. 2.6 Frontal view of a house dust mite, showing the chelicerae and the two foremost pairs of legs. The palps are concealed underneath the chelicerae, except for a small part of the left palp. *(Photo: K. Kuller.)*

Quite a number of mite species are harmful, either to human health, to stored goods, to agricultural plants, or to domestic animals. Because of the economic importance, many zoologists direct their attention entirely to this group. As a result, a special field has developed, known as "acarology." Acarologists from all over the world regularly get together in scientific meetings. The 8th International Congress of Acarology, held in Ceske Budejovice, Czechoslovakia (now the Czech republic) in August 1990, was announced with the subtitle: "A Modern Science." One of my colleagues at the time commented that this sounds apologetic. None considers acarology as a particularly modern branch of science. And why should we care? Old is not the same as obsolete. However, the microscopes used by acarologists are very modern and make an indispensable contribution to the science of acarology. They allow us to marvel at a plethora of forms and structures. The quality of the images is something that you must see to believe. And acarologists try to make some sense of what they see.

Many years ago I attended a symposium organized by the Dutch Entomological Society (NEV). One of the speakers began his presentation by saying that the taxonomy of the insects is based entirely on the integument, i.e., on the exoskeleton. It was meant as a sneer. He himself was using molecular techniques to clarify the taxonomic relations between insect species. But, a sneer it may have been, it is nevertheless true, not only for the taxonomy of insects but also for the taxonomy of mites. A taxonomic acarologist studies microscopic preparations of the outer skeleton, also called the integument, of the mite. To prepare it for microscopic examination, the mite is put in a drop of mounting fluid on a glass plate (micro slide) and covered by an even smaller glass plate (cover slip). Chemicals in the mounting fluid break down the organic macromolecules of the tissues so that, after some time, only the chitinous exoskeleton is left. A phase contrast microscope is used to study the structural details of the integument. And there is quite a lot to be seen. Besides the external shape of the mite's body, many of the internal organs and sense organs leave their marks on the integument (Table 2.1).

2.1 *Dermatophagoides* as an example of a mite

Dermatophagoides pteronyssinus and *Dermatophagoides farinae* are closely related species. Some aspects of their biology or morphology are accurately investigated in one of the two species but much less or not at all in the

Table 2.1 Microscopy techniques.

Light microscopy, resolution limit: 200 nm

Stereo or dissecting microscope
Suitable for observing live mites
Bright field microscopy
Illumination by a light source directly underneath the object, so that the
 background is bright
Dark field microscopy
Objects are illuminated by a light source that is not directly underneath it, so that the
 background is dark.
Phase contrast microscopy
Hyaline objects, devoid of any light absorbing pigments, such as the prepared
 exoskeleton of a mite, can be studied with the aid of these microscopes.
It exploits differences in refractive index inside the object.
It is the preferred type of light microscopy for the acarologist.
If one area of the object is in focus, other areas are slightly out of focus, so that
 photographic images directly made through a microscope are not satisfactory.
That is why acarologists prepare drawings: one drawing of the dorsal side and one of
 the ventral side.
Such a drawing is an interpretation; "not a mite but a merely a drawing of a mite."

Electron microscopy, resolution limit: 0.2 nm

SEM or Scanning Electron Microscopy
The object must be in high vacuum. (Mites can survive that.)
Its surface is scanned with a very narrow beam of electrons.
Features: sharp all over, no colors (unless artificially added)
TEM or Transmission Electron Microscopy
Tissues must be killed, fixed, treated with dyes, embedded, and sliced.

---

other one. Often it is taken for granted that there is no need to study a certain aspect equally well in both species because most probably there will be no differences. For instance, Brody et al. (1972) studied the anatomy of *D. farinae* using various microscopy techniques. Douglas and Hart (1989) studied transverse paraffin-wax sections of *D. pteronyssinus*, focusing mainly on the digestive tract and confirm that it is very similar in the two species. Dust mites are miniature animals, barely visible for the naked eye. An adult female of *D. pteronyssinus* measures about 400 μm. That is roughly 70 times the diameter of a eukaryotic cell. But unlike a eukaryotic cell, mites have legs, mouth parts, muscles, an intestinal tract, a nervous system, and more (Fig. 2.7). No blood vessels though. But larger arthropods have no closed system of blood vessels either. The internal organs are

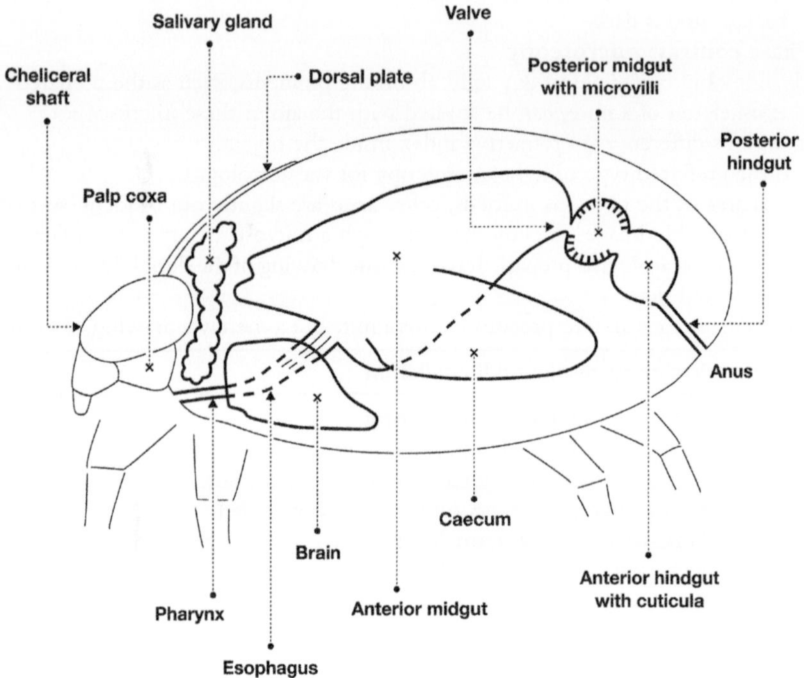

Fig. 2.7 Schematic presentation of the internal organs of an adult female house dust mite. *(Based on a study published by Brody, A.R., McGrath, J.C., Wharton, G.W., 1972. Dermatophagoides farinae, the digestive system. N.Y. Entomol. Soc. 80, 152–177. Artwork by Fanny de Boer.)*

embedded in a fluid that is called "the hemolymph." Transport of chemicals such as oxygen, CO_2, and nutrients depends on diffusion and on currents in the hemolymph. The hydrostatic pressure in the hemolymph is believed to be important for locomotion because this hemolymphatic pressure may serve as an antagonist of the muscles that bend the legs. Dehydrated dust mites sit still. It has been suggested that they could not move, even if they wanted to, because the hydrostatic pressure in the hemolymph is too low. Also the forward extension of the mouth parts (gnathosoma) is believed to be driven by this pressure, as well as the protrusion of the penis and anal suckers of males during copulation and of the ovipositor of females during egg deposition. The integumentary "shields," such as the prodorsal shield (Fig. 2.8e) of dust mites and other *Oribatida*, may be used to exert pressure on the hemolymph. Therefore these shields may be, indirectly, of crucial importance for locomotion.

The food of dust mites consists of particles of organic material, which they find in the house dust habitat, carpets, rugs, mattresses, and upholstery. The food particles are ingested using the palps and the chelicerae. Brody et al. (1972) present a SEM picture where it can be seen that the right chelicera of *D. farinae* is extended while the left one is retracted, showing the agility of these appendages. Each one of the chelicerae is equipped with a movable digit. The esophagus is star shaped in cross section so it can be widened to allow the passage of relatively large food particles. It is inferred that particles over 20 µm in diameter can be accommodated. A peculiar anatomical feature of mites is that the esophagus passes through the brain. In the more posterior part of the alimentary canal, balls of food are formed, each one surrounded by a so-called peritrophic membrane. Protection of the epithelium of the midgut has been suggested as a function of this membrane. *Dermatophagoides farinae* has no Malpighian tubules. The closely related *D. pteronyssinus* presumably has no Malpighian tubules either, though other mites and insects do. The function of the Malpighian tubules is to sequester metabolic waste products and to discharge these into the gut lumen. This function must have been taken over by other tissues in the alimentary canal, because it is well known that the fecal pellets of dust mites are full of the nitrogenous waste product guanine. One fecal pellet, as it is excreted by a dust mite, is composed of three to five food balls.

Hairs are very prominent on the body of a mite. However, both the structure and the function of a "hair" are totally different from the hairs of mammals. An acarologist refers to a hair as a "seta" (plural "setae"). Setae are sensory organs innervated by a basal nerve cell. Often they are

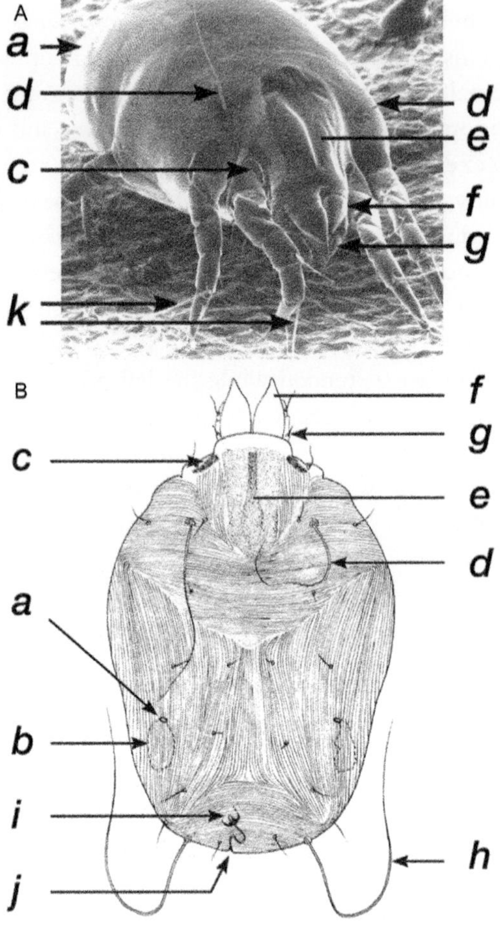
A
a
d
c
k
d
e
f
g
B
c
a
b
i
j
f
g
e
d
h

Fig. 2.8 (A) SEM picture of an adult female *D. pteronyssinus*. (B) Drawing based on a microscopic image of a mounted adult female *D. pteronyssinus*. Dorsal side. *(Artwork by the author using a drawing made by Fain, 1966.)*

a. orifice of opisthonotal gland

Function: see *b*

b. opisthonotal gland, also called opisthosomatic, latero-abdominal, or oil gland. A single-celled gland, characteristic for some oribatid taxa.

Function: not known with certainty, but the production of pheromones is a likely possibility.

c. supra coxal sclerite, with, in the center, the orifice of the corresponding supra coxal gland.

Function: It is hypothesized that the supra coxal glands have a key role in the process of water vapor uptake.

d. external scapular seta (sc e)

e. prodorsal shield (stippled in the drawing below).

Function: shields are believed to be important for attachment of muscles, exerting pressure on the hemolymph

f. chelicera

g. palp

h. internal sacral seta (sa I) (straight and directed backward in living mites)

i. sclerotized part of the receptaculum seminis, an internal organ, not visible in SEM

Function: storage of sperm

j. bursa copulatrix, connected with the receptaculum seminis through a sclerotized duct called the ductus receptaculi.

Function: copulation and transport of sperm to the receptaculum seminis

k. solenidion φ on the tibia of legs I and II (not shown in the drawing)

mechanoreceptors but also chemoreceptors, thermoreceptors, hygro-receptors, and photoreceptors have been identified in mites. Morphologically several different types of setae have been distinguished and named. Francois Grandjean (1882–1975), one of the "fathers of acarology," examined microscopic preparations of mites using polarized light and he found that some setae contain an optically active (anisotropic) material called "actinopilin" whereas other setae are optically inactive. Moreover, he found that two groups of mite species can be distinguished: one, the "*Anactinotrichida*," having no actinopilin containing setae and another one, the "*Actinotrichida*," of which the members do have actinopilin-containing setae. In the classification followed by Krantz and Walter (2009), the two groups are considered as two super orders that together make up the subclass *Acari*. One has the name *Parasitiformes* (*Anactinotrichida*) and the other one is called *Acariformes* (*Actinotrichida*). Dust mites are in the latter group. But not all the setae of dust mites contain actinopilin. One type of seta, called "solenidion" (plural: "solenidia"), has a protoplasmic core and no actinopilin. Dust mites and other members of the suborder *Oribatida* have a whip-like solenidion (φ) on the dorsal side of the first and second pair of legs (Figs. 2.8 and 2.9). These solenidia are very conspicuous, arising from the tibia, and extend beyond the end of the tarsus. Living mites project these solenidia straightforward so that they occupy a good position to act as tactile organs.

Table 2.2 presents how the taxonomic position of dust mites is based on morphological characters. Dust mites are representatives of a distinguished subgroup of mites, namely the cohort Astigmatina ("Astigs" for their intimi). As the name implies Astigmatina have no stigmata, i.e. no respiratory structures, so that gas exchange has to go through the skin. Many of these mites are very small ectoparasites of birds or mammals, spending their entire life on or in the skin, the fur, or the plumage of their host. Other species live only part of their life as parasites, in particular during the deutonymphal stage of their ontogeny (see Section 7). But dust mites are free living. Their closest relatives are nidicoles, i.e. live in the nests of birds or mammals. The relations between various species of the Astigmatina were studied by Klimov and Oconnor (2013) on the basis of the nucleotide sequences of five genes. The interpretation of such data is the domain of specialists, but one of their conclusions is too interesting not to mention it. They say that, most likely, the dust mites and their

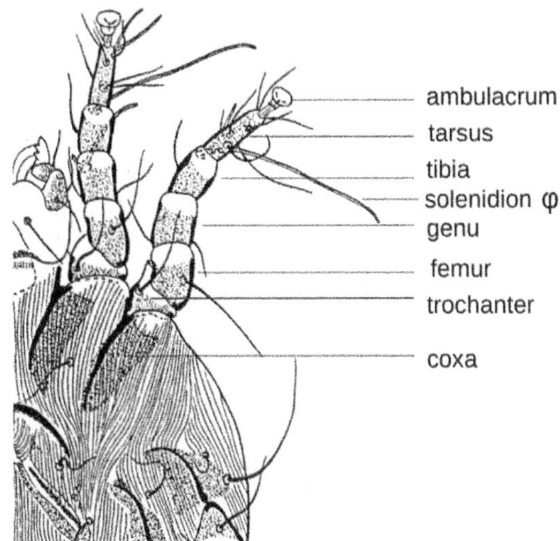

Fig. 2.9 *Dermatophagoides pteronyssinus*, legs 1 and 2, female, ventral view. Note that the solenidion φ arises from the dorsal side of the tibia of both legs. *(Artwork by the author using a drawing made by Fain, 1966.)*

Table 2.2 The systematic position of *Dermatophagoides pteronyssinus*.

Taxon	Hallmark traits
Kingdom: *Animalia*	Heterotrophic feeding
Subdivided into 33 phyla	Metabolism results in the production of nitrogenous waste which has to be excreted
Arthropoda	External chitinous skeleton
	Segmented body
Considered by Krantz and Walter as an unranked assemblage, but regarded by many authors as a phylum	Jointed legs
	A hemocoel (no closed network of blood vessels)
	Ventral nerve cord
	Alimentary canal
	Striated muscles
	(often) Malpighian tubules
	Separate males and females (with only a few exceptions)
Phylum: *Chelicerata*	No antennae, no mandibles, no compound eyes
Treated by many authors as a "subphylum" of the "phylum" *Arthropoda*	Body divided into a prosoma (bearing the appendages) and an opisthosoma (abdomen)

Continued

Table 2.2 The systematic position of *Dermatophagoides pteronyssinus*—cont'd

Taxon	Hallmark traits
Besides the *Chelicerata* there are five more phyla within the *Arthropoda* such as the *Crustacea* (crabs, lobsters, etc.) and the *Uniramia* (including the Insects)	One pair of preoral appendages (chelicerae), functioning as mouth parts
Class: *Arachnida*	Five pairs of postoral appendages comprising one pair of pedipalps and four pairs of walking legs
	All living arachnids are basically terrestrial (a few adapted secondarily
Besides the *Arachnida* there are two more classes within the *Chelicerata*. The *Arachnida* is the biggest of the three	to an aquatic lifestyle)
	Guanine is the most important nitrogenous waste product
	The excretory organs are coxal glands and/or Malpighian tubules
Subclass: *Acari (Acarina)*	Lack of spinnerets
	The body is unsegmented or indistinctly so
Webb et al. (1978) consider the *Acari* as an order within the class *Arachnida*, besides ten other orders such as:	Larvae have six legs and subsequent stages have eight legs
Aranea (spiders)	
Scorpiones (scorpions)	
Opiliones (harvestmen)	
Superorder: *Acariformes* (*Actinotrichida*)	Coxae I–IV fused to the body wall
	Tarsi without fissures
	Most sensory setae contain actinopilin that exhibits birefringence in polarized light (hence the designation *Actinotrichida*)
Besides the *Acariformes* there is one other superorder within the *Acari*, namely the *Parasitiformes (Anactinotrichida)*	
Order: *Sarcoptiformes*	Chelicerae typically chelate (pincer-like)
Besides the *Sarcoptiformes* there is one other order within the *Acariformes*, namely the *Trombidiformes*	
Some common inhabitants of house dust (members of the families *Tarsonemidae* and *Cheyletidae*) belong to the *Trombidiformes*	
Suborder: *Oribatida*	Mostly particle-feeding saprophages and mycophages
Besides the *Oribatida* there is one other suborder within the *Sarcoptiformes*, namely the *Endeostigmata*	The life cycle typically includes a calyptostatic prelarva, a mobile hexapod larva, three nymphs (proto-, deuto-, and tritonymph), and an adult.

Table 2.2 The systematic position of *Dermatophagoides pteronyssinus*—cont'd

Taxon	Hallmark traits
	Some oribatid taxa have opisthonotal glands
Supercohort: *Desmonomatides* (*Desmonomata*)	Possess, with only few exceptions, a pair of opisthonotal glands
Besides the *Desmonomatides* there are four more supercohorts within the *Oribatida*	
Cohort: *Astigmatina* (*Astigmata*)	Lack of specialized respiratory structures
	Transdehiscent ecdysis
Besides the *Astigmatina* there are two more cohorts within the *Desmonomatides*	Sperm is introduced into the bursa copulatrix of the female, which is a special entrance separate from the ovipore
	A facultative, heteromorphic deutonymph (hypopus) occurs in many species
	Chelicerae prominent, palps not so
Psoroptidia	Loss of the deutonymph from the life cycle so that there is no hypopus
"… an unranked but hypothetically monophyletic group that includes most of the permanent astigmatine symbionts of vertebrates …" (Krantz and Walter, 2009)	Striated cuticle
Superfamily: *Analgoidea*	Mostly paraphages and parasites of birds
Besides the *Analgoidea* there are nine other superfamilies within the *Astigmatina*. Two of these, the *Glycyphagoidea* and the *Acaroidea*, have representatives living in house dust	
Family: *Pyroglyphidae*	Mostly inhabitants of the nests of birds and mammals, feeding on organic debris, but some genera are ectoparasites of birds
This family was grouped by OConnor (1982) with the *Ptyssalgidae* and the *Turbinoptidae* into a separate superfamily, the *Pyroglyphidea*. Besides these three there are 17 more families within the *Analgoidea*. One of these is the *Epidermoptidae*. Some authors assign *Dermatophagoides* to this family.	Retroconjugate copulation (i.e. heads facing in opposite directions)
	Males with anal suckers

Continued

Table 2.2 The systematic position of *Dermatophagoides pteronyssinus*—cont'd

Taxon	Hallmark traits
Genus: *Dermatophagoides*	Descriptions and keys to species in: Hughes (1976)
Besides *Dermatophagoides* there are 18 other genera within the *Pyroglyphidae*	Fain et al. (1990) Colloff and Spieksma (1992) Colloff (2009)
Species: *Dermatophagoides pteronyssinus*	
Besides *D. pteronyssinus* at least ten more species in the genus *Dermatophagoides* have been described	

For the classification of the Acari, Krantz and Walter (2009) were followed. For higher classifications Webb et al. (1978) were followed.

nidicolous relatives evolved from ancestors that were ectoparasites of birds, living in the plumage of their hosts. That is a violation of Dollo's law, because Dollo's law holds that evolution toward a parasitic lifestyle is irreversible.

2.2 Sampling houses for dust mites and setting up cultures

Dust samples from floors, furniture, and beds can be taken with an ordinary vacuum cleaner. There is no standardized procedure that is satisfactory in every respect. A portion of the vacuumed dust is suspended in a liquid and transferred to a petri dish. For the 'flotation method' a high density liquid is used so that the mites tend to float near the surface and can be picked up and transferred to a drop of mounting fluid on a micro-slide for microscopic examination. A better chance to recover every last mite from a (sub)sample can be achieved when a very thin layer of suspension is made in the petri dish and examined with dark field illumination. When a mite is complete and undamaged it is taken for granted that it was still alive at the time when it was trapped by the vacuum cleaner (see also Appendices A and H).

A very different method is the so-called mobility test. With this method only live mites are captured, which is often an advantage. A piece of sticky tape is stuck to, for instance, a carpet. After some time the tape is pulled off and the sticky side of it is examined through a dissecting microscope (Figs. 2.10B and 2.11). The sticky tape method can be modified into the

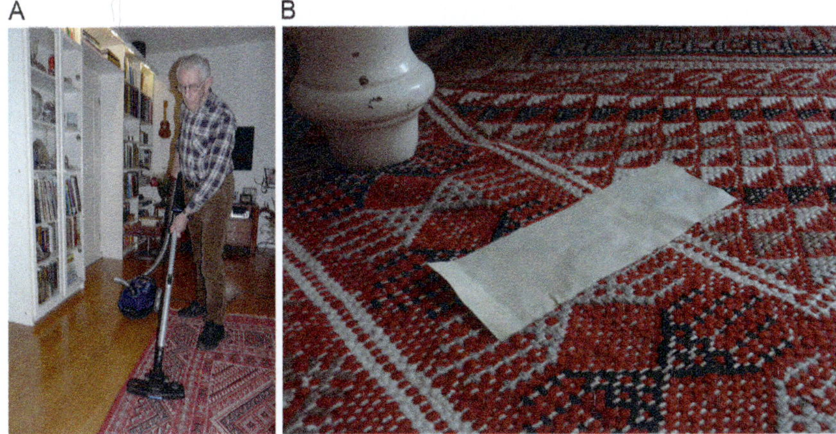

Fig. 2.10 (A) Taking a dust sample. (B) Using sticky tape for the "Mobility test."

Fig. 2.11 Dust mites (*D. pteronyssinus*) captured on sticky tape by the "Heat escape method." The egg in the middle of the picture must have been produced by a female who got detached from the sticky surface when the tape was removed from the substrate (an artificially infested carpet fragment).

"heat escape method," but then the carpet must be sacrificed. Fragments of the carpet are placed on a slide warmer and heated from below. The mites that stick to the tape can be counted but they cannot be transferred to a micro-slide to determine the species. Eggs and immobile stages of mites are not captured, which is another disadvantage. A small sample taken with a vacuum cleaner, in addition to the heat escape sampling, may partly make up for these disadvantages. Also for experimental work, using cultured mites, the heat escape method offers good possibilities. Details of statistical evaluations are given in Appendix A.

Hirschmann (1986, 1988) describes (in German) a very simple way to collect dust mites. Just put a sheet of paper on the infested carpet or mattress. After some time mites will be walking on the paper, both on the underside and the upper side. The captured mites are unharmed. Therefore this may be the preferred method for certain purposes, such as setting up cultures or to examine whether an adult female is inseminated and able to lay eggs. Hirschmann reported to have collected the species *Lepidoglyphus destructor* and *Dermatophagoides farinae* but no *D. pteronyssinus*. I fear that this reflects a difference in behavior. In cultures *D. pteronyssinus* is also more reluctant than *D. farinae* to walk on the walls of the culture vials, as noted by myself and others.

In yet another article, also in German, Hirschmann and Kemnitzer (1988) describe a procedure whereby dust out of the bag of a vacuum cleaner is sieved and spread on a sheet of paper. The mites can be teased out of the dust using a lamp. Whether the light or the heat from the lamp is teasing them is an interesting question and amenable for experimental examination.

Dust mites can be cultured quite well. Dry organic material, for instance a mixture of dried yeast and dried *Daphnia* (for sale as food for aquarium fish), ground to small particles can be offered as food.

References

Brody, A.R., McGrath, J.C., Wharton, G.W., 1972. Dermatophagoides farinae, the digestive system. N. Y. Entomol. Soc. 80, 152–177.

Colloff, M.D., 2009. Dust Mites. CSIRO/Springer Science, ISBN: 978-90-481-2223-3, p. 583.

Colloff, M.D., Spieksma, F.Th.M., 1992. Pictorial keys for the identification of domestic mites. Clin. Exp. Allergy 22, 823–830.

Douglas, A.E., Hart, B.J., 1989. The significance of the fungus Aspergillus peniciloides to the house dust mite Dermatophagoides pteronyssinus. Symbiosis 7, 105–116.

Fain, A., Guerin, B., Hart, B.J., 1990. In: Guerin, B. (Ed.), Mites and Allergic Disease. Allerbio, Varenne en Argonne, France. 190 pp.

Fain, A., 1966. Acarologia. Nouvelle description de Dermatophagoides pteronyssinus (Trouessart, 1897) Importance de cet acarien en pathologie humaine (Psoroptidae: Sarcoptiformes). 8, 302–327.

Hirschmann, W., 1986. Einfaches Verfahren zum Nachweis von lebenden Hausstaubilben. Der praktische Schädling Bekämpfer 38, 130–131.

Hirschmann, W., 1988. Nachweis lebender Hausstaubmilben. Der praktische Schädling Bekämpfer 2 88, 18–20.

Hirschmann, W., Kemnitzer, F., 1988. Hausstaubilben. Mikrokosmos 77, 117–122.

Hughes, A.M., 1976. Mites of Stored Food and Houses: Technical Bulletin 9. Ministry of Agriculture, Fisheries and Food, London. ISBN: 0 11 240909 1.

Klimov, P.B., Oconnor, B., 2013. Is permanent parasitism reversible?—Critical evidence from early evolution of house dust mites. Syst. Biol. 62, 411–423.

Krantz, G.W., Walter, D.E., 2009. A Manual of Acarology, third ed. Texas Tech University Press. ISBN 978-0-89672-620-8.

OConnor, B.M., 1982. Evolutionary ecology of astimatid mites. Annu. Rev. Entomol 27, 385–409.

Webb, J.E., Wallwork, J.A., Elgood, J.H., 1978. Guide to Invertebrate Animals, second ed. McMillan Press Ltd.

Table 3.1 The four species comprising a typical house dust mite fauna as it can be found everywhere on earth.

	% of all mites	Infested houses
Pyroglyphidae		
Dermatophagoides pteronyssinus	76	15 (100%)
Dermatophagoides farinae or *microceras*	6	8 (53%)
Euroglyphus maynei	13	12 (80%)
Cheyletidae		
Cheyletus sp.	3	15 (100%)
Miscellaneous mite species	2	

This example shows the mites collected in 15 houses in Santiago de Compostela, Spain, which were studied by Agratorres et al. (1999). Details in Appendix B.

Table 3.2 Taxa that are found less often in house dust but occasionally in high numbers.

Taxon	% of all 55 locations where this taxon was found	
Glycyphagidae	73	
Blomia sp.	29	mostly *B. tropicalis*
Glycyphagus sp.	35	mostly *G. domesticus*
Chortoglyphidae	18	
Chortoglyphus sp.	18	mostly *C. arcuatus*
Pyroglyphidae	100	
Malayoglyphus sp.	13	mostly *M. intermedius*
Sturnophagoides sp.	4	in particular *S. brasiliensis*
Acaridae	51	
Tyrophagus sp.	35	mostly *T. putrescentiae*
Tarsonemidae	35	
Tarsonemus sp.	18	

Details in Appendix B.

CHAPTER THREE

House dust faunas around the world

Appendix B lists 32 studies concerning the mite fauna in house dust, published between 1970 and 1999. Together these inform us of the dust mite fauna in 55 locations. Every continent, except Antarctica, is represented: 15 of the locations were in Europe, 13 in North America, 11 in South America, 11 in Asia, 3 in sub-Saharan Africa, and 2 in Australia. Only 13 of these locations are in the tropics (latitude <23 degrees). Norway and Finland have the most northern locations. The European house dust mite, *Dermatophagoides pteronyssinus*, is the only species that was found in all these places, although not necessarily in every dwelling. In only very few houses it was absent and then it may concern a house where no mites were found at all. Very often it is also the dominant species, outnumbering all other species in a dust sample. Table 3.1. gives just one example of a house dust fauna. This one was recorded in Spain, but house dust faunas with this assemblage of species are found commonly all over the world. Three species belonging to the *Pyroglyphidae* occur in fairly large numbers, namely *D. pteronyssinus*, *D. farinae*, *Euroglyphus maynei*, and very often there is also a member of the predatory *Cheyletidae*. No doubt each one of these species completes its entire life cycle within the house dust habitat and therefore may be considered a true house dust mite. The three species of *Pyroglyphidae* feed on organic debris. The *Cheyletidae* are predators that presumably attack the pyroglyphid mites and feed on them.

In most of the 55 locations listed in the Appendix, *D. pteronyssinus* was more numerous, often far more numerous, than *D. farinae*. But, for instance, in Nir David (Israel) and in Nagoya (Japan) the two species were equally common, and in Helsinki (Finland) 62% of all mites were *D. farinae*, whereas only 23% were *D. pteronyssinus*. On the other hand, in the two locations in Australia that were thoroughly studied by Colloff, *D. farinae* was entirely absent.

Within one location the composition of the acarofauna in individual houses may be quite different. For instance, Arlian et al. (1992) studied the house dust fauna of 48 homes in Cincinnati, Ohio, USA. Ten of these houses were inhabited by only *D. farinae*, two had only *D. pteronyssinus,* and 36 were coinhabited by both species.

House Dust Mites
https://doi.org/10.1016/B978-0-443-19111-4.00004-6

E. maynei is also quite common. It is found in most of the locations studied, ranging from the tropics to Finland and may comprise up to 47% of the population (Bunbury, Australia).

The predatory *Cheyletidae* are one level higher in the food pyramid. They are less numerous than the pyroglyphids, but in the majority of the studied locations one or more representatives of the *Cheyletidae* were found. *Cheyletus eruditus* was found most frequently, but three other *Cheyletus* species and eight *Cheyletidae* belonging to other genera were also recorded.

Cunnington (1971) described how a new species of dust mite that closely resembled *D. farinae* was discovered at the Pest Infestation Laboratory in Slough (UK). At this laboratory a culture of *D. farinae* of American origin was maintained and used since 1963. In 1966 a coworker of this laboratory isolated mites from house dust collected in London (UK) and set up a culture of it. The mites were initially identified as *D. farinae*, but comparison with the American strain revealed great differences in rates of growth and development. Hybridization experiments were carried out and the results of these showed "not only that each population was completely reproductively isolated from the other but that mutual attraction between opposite sexes of the different populations was also absent." A detailed micromorphological investigation was undertaken by D.A. Griffiths, who concluded that the British strain was a discrete species new to science and he proposed the name *Dermatophagoides microceras* sp.n. (Griffiths and Cunnington, 1971).

D. microceras can be distinguished morphologically from *D. farinae*, but the differences are trifling. Even in studies that were done later than 1971 some of the mites identified as *D. farinae* may in fact be *D. microceras*. Sometimes the designation "*D. farinae*-group" is used to indicate that the species concerned may be either *D. farinae* or *D. microceras*. Careful, time-consuming examination of the tarsi at a high magnification requiring oil immersion is necessary to distinguish the two species. Not every investigator bothers to do that, but some do. And remarkable differences in the distribution of the two species were found. In a survey of 108 houses in seven locations in Norway, ranging from 59.15 to 64.56 degrees North Latitude, no *D. farinae* was found but *D. microceras* occurred in 62% of the houses (Mehl, 1998). In contrast, no *D. microceras* was found among the thousands of mites that were collected in a survey of 252 houses in eight locations in the United States. This was explicitly stated, which means that the authors did not just ignore it but consciously examined every *D. farinae* for *microceras* characters (Arlian et al., 1992).

To my mind the discovery of *D. microceras* is somewhat disturbing. How many more sibling species exist that can be brought to light only by hybridization experiments?

In Table 3.2. species are listed that were found occasionally in high numbers in samples of house dust, but not so often and not so widespread as the three pyroglyphids *D. pteronyssinus*, *D. farinae*, and *E. maynei*.

Of the rarer species, *Blomia tropicalis* (*Glycyphagidae*) is the only one that may locally rival *D. pteronyssinus* for the first position on the abundance scale, but only in the warmer parts of the world. It was found in each one of 72 houses in several locations in Brazil, mostly around Sao Paulo. Almost 80% of the 2198 mites found were *B. tropicalis*. Only 4% were *D. pteronyssinus* and all pyroglyphids together made up only 7%. *Blomia* sp. was reported from many other locations in the tropics and the subtropics, but nowhere else in such high numbers.

Glycyphagus sp., in particular *Glycyphagus domesticus*, seems to be more common in the colder regions of the world. In his book entitled *Dust Mites*, published in 2009, Colloff (2009) refers to his own study of the dust mite faunas in Glasgow, Scotland, saying that: "*Glycyphagus domesticus* is present in 14% of homes but is more abundant than *E. maynei* (which is present in 32% of homes), suggesting that large populations of this mite develop in relatively few homes. There must be something different about these homes to support high populations of this species (in fact, they tend to be very damp)." In Norway *G. domesticus* was found "in high numbers in humid homes" according to Mehl (1998).

In Saladito (Colombia) 54% of all 880 mites found belonged to the family *Chortoglyphidae* (Charlet et al., 1979). Representatives of this family were found in seven other locations, ranging from Colombia to Switzerland, but never in such high numbers.

Two other pyroglyphid species have also been found in house dust, namely *Malayoglyphus intermedius* and *Sturnophagoides brasiliensis*. In their paper on the dust mites of Colombia, Charlet et al. (1979) say that: "… during the period of sampling it (*D. pteronyssinus*) was also the most abundant on both the floor and in mattresses, except for Buenaventura where *M. intermedius* was more abundant. …" Also *Tyrophagus* sp. (*Acaridae*) and *Tarsonemus* sp. (*Tarsonemidae*) were not uncommon.

Hurtado and Parini (1987) say about the dust mites in Caracas (Venezuela): "The genera *Tyrophagus* and *Tarsonemus* were infrequent but occasionally large contributors to the bedding dust fauna."

One other example: House dust faunas in the prefecture Saitama in Japan were studied by Takaoka and Okada (1984) who wrote: "Mites that were found in substantial numbers belonged to the taxa *D. pteronyssinus*, *D. farinae* and the *Tarsonemidae*."

When we focus on mite families rather than mite species, the following conclusions can be drawn: Six mite families occur in house dust worldwide (Tables 3.1 and 3.2). The *Pyroglyphidae* are best represented, with five species, three of which are very common. Most species of this family are inhabitants of the nests of birds or mammals, where they feed on organic debris, but a few are obligate parasites of birds. The predatory *Cheyletidae* are represented by 12 species that are commonly encountered in house dust but never in high numbers. The *Glycyphagidae*, which is also a family of inhabitants of the nests of vertebrates, are represented by two genera, *Blomia* and *Glycyphagus*. Little is known of the natural ecology of these mites. The *Chortoglyphidae* are by some authors regarded as a subfamily of the *Glycyphagidae*.

Tyrophagus belongs to the family *Acaridae*. The *Acaridae* are "ecologically diverse" (Krantz and Walter, 2009, p. 580). *Tyrophagus putrescentiae* is "perhaps the most ubiquitous mite, that lives in many peri-domestic situations and also on plant tissue and is a common pest in laboratory insect and fungus cultures" (Krantz and Walter, 2009, p. 581; Hughes, 1976). *Tarsonemus* belongs to yet another family, the *Tarsonemidae*. Species of this family are either phytophagous, algivorous, or fungivorous. When they occur massively in house dust they may perhaps feed on fungi.

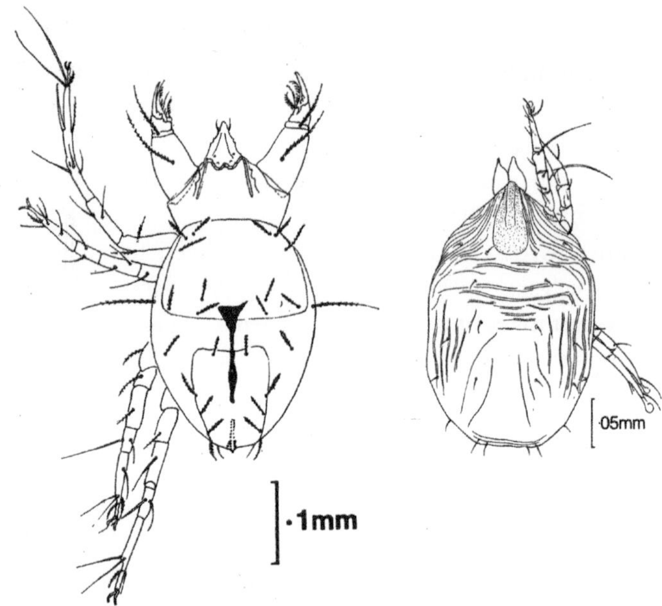

Fig. 3.1 Two types of mites that are almost as ubiquitous in house dust as the *Dermatophagoides* species. Left: *Cheyletus eruditus*, a member of the predatory family *Cheyletidae*, of which several species occur in house dust. Note the large palps that are used to capture prey. Right: *Euroglyphus maynei* like *Dermatophagoides* is a member of the *Pyroglyphidae*. It is of about the same size as *Dermatophagoides*, but it does not have the elongated scapular setae. Note also the coarse, irregular striations and the pointed front (tegmen) of the dorsum. *(From Hughes, A.M., 1976. Mites of Stored Food and Houses: Technical Bulletin 9. Ministry of Agriculture, Fisheries and Food, London. ISBN: 0-11-240909.1.)*

Thus, a typical house dust fauna consists of one, two, or three species of the *Pyroglyphidae* (order *Sarcoptiformes*), feeding on detritus, and one 'hunting mite' belonging to the *Cheyletidae* (order *Trombidiformes*). Specific locations or individual houses may depart from this species composition. In the tropics *Blomia*, a representative of the *Glycyphagidae*, may be the dominant mite, and not one of the *Pyroglyphidae* (Figs. 3.1 and 3.2).

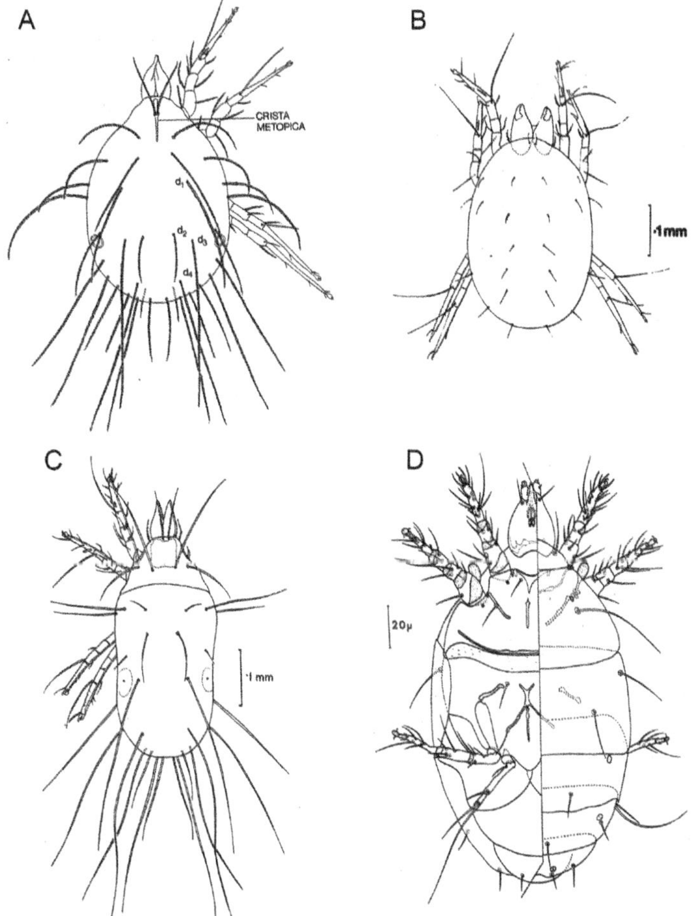

Fig. 3.2 Four examples of species, representing four mite families, that are occasionally found in large numbers in house dust.

A. *Glycyphagus domesticus (Glycyphagidae).* Male, dorsal view.

B. *Chortoglyphus arcuatus (Chortoglyphidae).* Male, dorsal view.

C. *Tyrophagus putrescentiae (Acaridae).* Male, dorsal view.

D. *Tarsonemus granarius (Tarsonemidae).* Female, left side: ventral view, right side: dorsal view.

(From Hughes, A.M., 1976. Mites of Stored Food and Houses: Technical Bulletin 9. Ministry of Agriculture, Fisheries and Food, London. ISBN: 0-11-240909.1.)

References

Agratorres, J.M., Pareira-Lorenzo, A., Fernandez-Fernandez, I., 1999. Population dynamics of house dust mites in Santiago de Compostela (Galicia, Spain). Acarologia 40, 59–63.

Arlian, L.G., Bernstein, D., Bernstein, I.L., Friedman, S., Grant, A., Lieberman, P., Lopez, M., Mezger, J., Platts-Mills, T., Schatz, M., Spector, S., Wasserman, S.I., Zeiger, R.S., 1992. Prevalence of dust mites in the homes of people with asthma living in eight different geographic areas of the U.S. J. Allergy Clin. Immunol. 90, 292–300.

Charlet, L.D., Mula, M.S., Sanchez-Medina, M., Rayes, M.A., 1979. Species composition and population trends of mites in various climatic zones of Colombia. J. Asthma Res. 16, 131–148.

Colloff, M.D., 2009. Dust Mites. CSIRO/Springer Science. 583 pp. ISBN: 978-90-481-2223-3.

Cunnington, A.M., 1971. House dust mites and respiratory allergy: a note on the identity of *Dermatophagoides farinae*. Clin. Allergy 1, 447–449.

Griffiths, D.A., Cunnington, 1971. *Dermatophagoides microceras* sp.n.: a description and comparison with its sibling species, *D. farinae* Hughes, 1961. J. Stored Prod. Res. 7, 1–14.

Hughes, A.M., 1976. Mites of Stored Food and Houses: Technical Bulletin 9. Ministry of Agriculture, Fisheries and Food, London. ISBN: 0-11-240909.1.

Hurtado, I., Parini, M., 1987. House dust mites in Caracas, Venezuela. Ann. Allergy 59, 128–130.

Krantz, G.W., Walter, D.E., 2009. A Manual of Acarology, third ed. Texas Tech University Press.

Mehl, R., 1998. Occurrence of mites in Norway and the rest of Scandinavia. Allergy 53 (Suppl. 48), 28–35.

Takaoka, M., Okada, S., 1984. Ecological studies on the house dust mites in Saitama prefecture Japan; seasonal occurrences of the pyroglyphid mites in house dust in 1981 to 1982. Jpn. J. San. Zool. 35, 129–137.

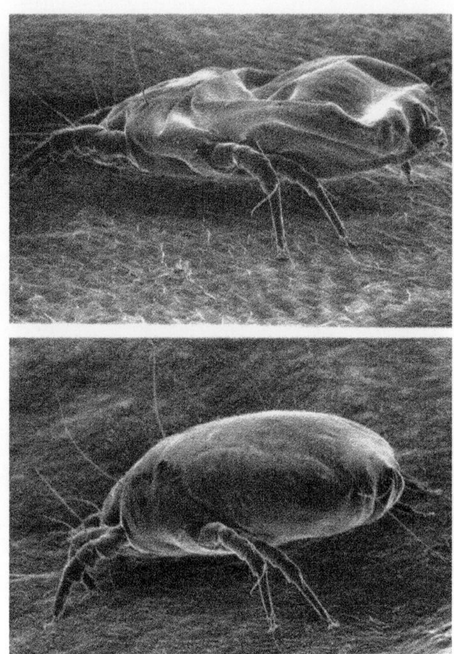

Fig. 4.1 Two house dust mites, *Dermatophagoides pteronyssinus*. Top: partly dehydrated, but alive. Bottom: a well-watered mite.

CHAPTER FOUR

Water junks

Once I talked with employees of the community health service (GGD) in the harbor city of Rotterdam (Netherlands) about their plans for a "healthy home project," houses especially designed for allergy patients. These people in their profession not only deal with regular health problems but also with drug addicts. After I explained how important the humidity is for dust mites, one of them reacted by saying: "So dust mites are water junks." I found the designation amusing but also indicative of a commonly held misconception. A dust mite was believed to be constantly engaged in an obsessive search for water, like a drug addict is for his drug. The importance of water vapor for the survival of dust mites is frequently emphasized. This led many people to associate dust mites with dampness. The notion that house dust mites are creatures that like warm and damp places is widely held. But there is something of a contradiction in this association with dampness because dust mites must be ranked among the most xerophilic creatures on earth. The house dust habitat is comparable with the most arid deserts. No liquid water is available at all—zero. So from an ecological perspective dust mites should be associated with aridity rather than with dampness. They are exceptionally well equipped to live in an extremely arid habitat. I have seen striking manifestations of this. For one of my investigations I obtained a rug from an upstairs bedroom in a private home in Amsterdam. Like the bedroom itself the rug was very dry and it also looked very clean. No sign of dampness; rather the opposite. But it turned out to harbor a large and thriving population of very healthy dust mites.

Very few organisms can survive in the house dust habitat. Dust mites can. Flea larvae can. Certain fungi can. But about the role of these last mentioned organisms we know very little.

Like all living creatures the bodies of dust mites are made up of cells consisting of water with solutes, enveloped in a membrane. So, like every other organism, dust mites need a source of water. The water vapor in the air that surrounds them may be the only source available.

House Dust Mites
https://doi.org/10.1016/B978-0-443-19111-4.00003-4

4.1 Food and metabolic water

As human beings we obtain most of our body water by drinking. Some water is also obtained by eating, because most solid foods contain a quantity of free water. In addition to that, the oxidation of carbohydrates, proteins, and fats invariably has CO_2 and H_2O among the end products. This so-called metabolic water could be the principal source of body water for some animal species. For instance, wood-boring beetle larvae have access to lots of combustible material all the time. In contrast, dust mites usually have a limited food supply and therefore food is an unreliable source of water. Arlian (1977) compared the feeding behavior in dry air with the feeding behavior in moist air, both for *Dermatophagoides farinae* and *Dermatophagoides pteronyssinus*. In extremely dry air, mites have very little interest in food. They were observed "moving slowly about the surface or they clustered together in a groove at either end of the cage." It has been suggested that in desiccating conditions movements of mites may be hampered or even rendered impossible because the turgor of the mite's hemolymph decreases due to water loss. And the turgor is important for locomotion, as explained in Section 4.2. Anyway, it is evident that the water obtained from food cannot be an important source of body water for dust mites.

4.2 The critical equilibrium (air)humidity (CEH)

(Technical explanations are given in Appendix D)

The only way for a house dust mite to obtain enough body water is by taking up water vapor from the air. Of course, there must be a certain amount of water vapor in the air to be able to do that. The minimum concentration that is required is called the Critical Equilibrium Humidity (CEH). For instance, at 20°C the CEH of *D. pteronyssinus* is somewhere between 58% and 60% relative humidity (RH). When the RH of the surrounding air is below this value a dust mite will gradually lose body water, shrink, and eventually die. Arlian et al. (1999) found that eggs of *D. farinae* will hatch when the RH is below the CEH, but the larvae will soon die.

For one of my early experiments I needed partially dehydrated dust mites. I confined mites individually in stainless steel cages, which were placed in a small container, in which the RH was kept at a very low level. I distinctly remember the first mite that I treated this way.

Inexperienced as I was, I kept it too long in dry air. The mite looked even more dehydrated than the one shown in Fig. 4.1. It resembled an old, shed plant leaf, thin, shriveled, and completely dried up. But the legs were moving slowly. Still alive but apparently moribund, I put it back into the cage and placed it in a container with high RH (75%). The next day I expected to find it dead. But when I looked I could hardly believe my eyes. The mite I saw was swollen like a balloon. Not only was it obviously well watered, it was also vividly walking about, looking as healthy as a house dust mite can be. Nothing reminded of the piteous creature that I saw the day before.

Arlian and Veselica (1981a, b) determined the CEH values of *D. farinae* at four different temperatures. The CEH value is usually expressed in %RH. The equivalent Absolute Humidity (AH) and Saturation Deficit can be calculated from it (Appendix C) and these values are presented in Table 4.1. In the earlier papers by Arlian and his mentor professor G.W. Wharton, the "water vapor activity (a_v)" is used as a measure of air humidity and the threshold value is called CEA (Critical Equilibrium *Activity*) instead of CEH. Water activity is a measure of humidity that is rarely used except in the food industry. It is very easy to recalculate water activity into % RH, namely by multiplying $\times 100$.

When the air humidity is above the CEH, the water mass in a mite's body develops toward an equilibrium level. The size of this equilibrium water mass appears to be independent of the ambient RH as long as it is above the CEH. Arlian and Wharton (1974) standardized *D. farinae* females at 75%RH. When these were transferred to either 85%, 95%, or 100% RH, the body mass remained unchanged. In this respect there seems to be a contrasting difference with another well-studied astigmatic mite, namely the "Flour mite," *Acarus siro* (Knülle, 1961, 1962, 1965). The CEH of these mites is between 70% and 75% RH at 22°C. Unlike *D. farinae* the equilibrium water content of this mite at humidities above the CEH is higher at higher ambient air humidities. Like hygroscopic materials the water content increased with increasing air humidity. So there is no standard equilibrium quantity of body water like in *D. farinae*. Observations of the behavior of *A. siro* showed that the mites seek places with high humidity and stay there to replenish their body water reserves. The elegant experiments and observations done by professor Knülle could and should also be done with dust mites. It is conceivable that, sometimes, the vertical RH gradient in a carpet spans the CEH of a dust mite, so that active migrations could help them survive.

Like dust mites *Acarus siro* belongs to the cohort *Astigmatina*, but to a different family; not one of the *Psoroptidia*. They lack the conspicuous corrugations of the cuticula, which may be relevant in connection with the way in which they maintain water balance.

Table 4.1 The critical equilibrium humidity (CEH) values of *Dermatophagoides farinae*.

	CEH		
Temperature (°C)	RH (%)	AH (g/kg)	Sat. def. (mbar)
15	~53 (45–55)	~6	~8
25	~58 (55–65)	~11	~13
30	~63 (55–65)	~16	~16
35	~70 (65–75)	~24	~17

AH, absolute humidity; *RH*, relative humidity; *Sat. def.*, saturation deficit.
Data from Arlian, L.G., Veselica, M.M., 1981a. Effect of temperature on the equilibrium body water mass in the mite *Dermatophagoides farinae*. Physiol. Zool. 54, 393–399; Arlian, L.G., Veselica, M.M., 1981b. Reevaluation of the humidity requirements of the house dust mite *Dermatophagoides farinae*. Med. Entomol. 18, 351–352.

4.3 The speed of water vapor uptake and loss

Fig. 4.2. shows the speed of water vapor uptake by partially dehydrated mites. Time is measured in hours. The next picture, Fig. 4.3., shows the speed of water loss when RH < CEH. Here time is measured in days. Clearly, water vapor uptake when RH > CEH proceeds much faster than water loss when RH < CEH.

It also matters how much the RH departs from the CEH. When RH is just a little above the CEH, the rate of uptake is significantly lower than when it is well above the CEH (de Boer, 2000). The CEH is a turning point. A little above the CEH mites can maintain water balance, but only just. When RH is well above the CEH, life is much easier for them. This is also apparent from the feeding rates. Arlian (1977) found that feeding rates and interest in food increase with increasing RH. Most likely, the reproduction rate is similarly influenced by the humidity of the surrounding air.

Differences between individual mites concerning the rate of water loss can be considerable, even when they are subjected to the same RH. Three mites subjected to 42%–44% RH lost between 31% and 47.5% of their initial water mass during 7 days. When the water loss continues at the same rate, it would take eight to 14 days until 50% of the initial amount of body water is lost. *D. farinae* has been shown to survive a loss of 52% of its original water mass (Arlian and Veselica, 1979).

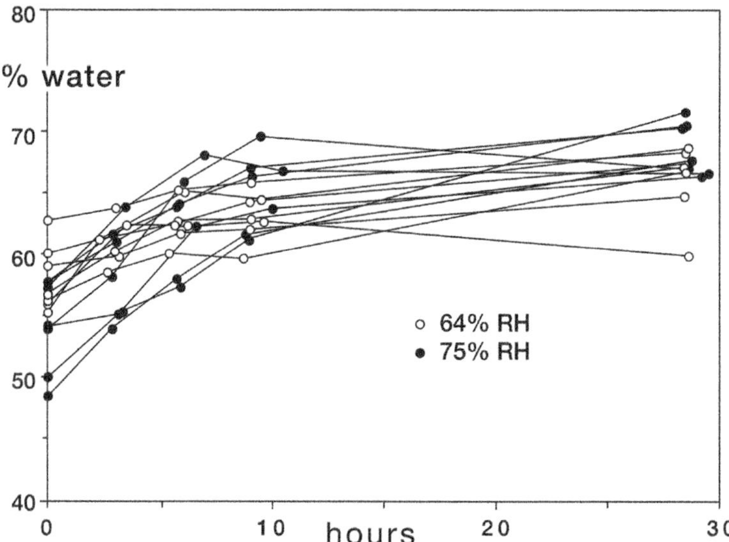

Fig. 4.2 The speed of water vapor uptake. Rehydration of partially dehydrated *Dermatophagoides pteronyssinus* at 18°C (de Boer, 2000).

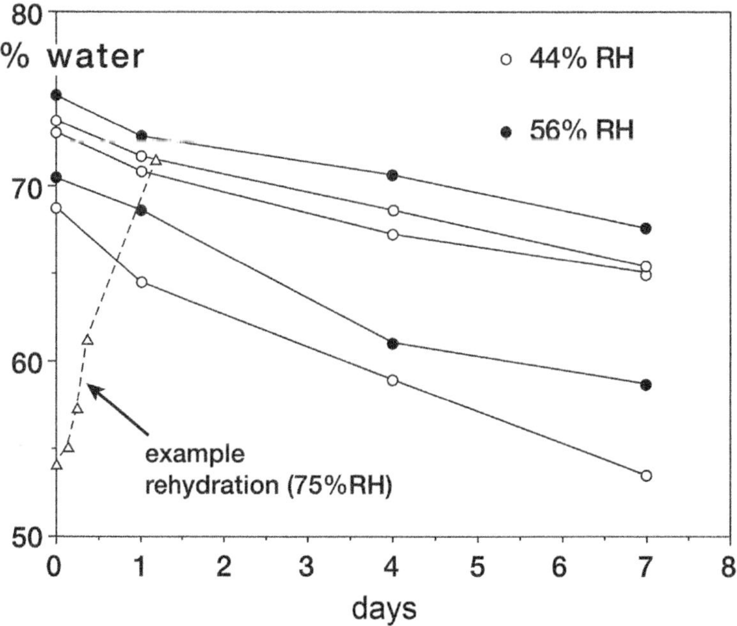

Fig. 4.3 The speed of water vapor loss. Dehydration of *Dermatophagoides pteronyssinus* at 18°C. Note that the timescale is different from Fig. 4.2. For easy comparison with the speed of water vapor uptake, one example of a rehydrating mite is also depicted (de Boer, 2000). *(Artwork by Fanny de Boer.)*

Often there are frequent fluctuations of the humidity in the habitats where house dust mites live. Situations may exist whereby the RH rises above the CEH during only a few hours every day (see Fig. 6.4). In order to study survival and reproduction in such circumstances, we built an apparatus to expose mites to daily cycles of alternating high and low RH. Stainless steel cages with one or more mites were placed in a rectangular chamber made of Perspex. An open side of the chamber is in connection with either a compartment with dry air (in contact with dry silica gel or a saturated salt solution) or a compartment with moist air. The air was stirred constantly. A pneumatic system, controlled by a time switch, slid the Perspex chamber with mite cages between compartments. A special chamber, designed to fit a temperature/RH probe, was used to measure RH inside the chamber.

For one series of experiments we placed the apparatus at a constant 16°C, a common temperature on the ground floor of Dutch houses in winter. The RH in the two compartments was kept at 36% (over saturated $MgCl_2$) and 76% (over saturated NaCl), respectively. The cages were filled with subcultures of our strain of *D. pteronyssinus*. An overview of some of the results is given in Table 4.2. The regimes *b* through *e* were an alternation of 36% and 76% RH, but mites subjected to regime *a* had a constant 44% RH (over saturated K_2CO_3). It is noteworthy that 1.5 h at 76% RH alternated with 22.5 h at 36% RH corresponds with an average 38.5% RH, i.e., less than 44%. Yet, at a constant 44% RH, mite numbers decreased significantly faster. This shows that it can be misleading to average the relative humidity over a period.

Table 4.2 Survival and reproduction of subcultures of *Dermatophagoides pteronyssinus* during a 10-week period at 16°C (de Boer et al., 1998).

Regime	Hours per day with moist (>CEH) air	Effect
a	0	Mites dry out It takes >10 weeks before all mites are dead
b	1½	Mites dry out It takes much longer than under Regime a, before all mites are dead
c	3	Mites do not dry out There is some reproduction
d	6	Faster reproduction than under Regime c
e	24	Yet faster reproduction than under Regime d

For these experiments the mites were subjected to the various regimes during a 10-week period. Ten weeks constitute a substantial part of the winter, and periods of extreme cold and dry weather rarely last that long in a maritime climate. Indeed, it has been shown that *D. pteronyssinus* can survive and remain active on the ground floor of Dutch houses throughout the winter (de Boer and Kuller, 1995).

Mites were also studied individually. Partially dehydrated, fasting mites were weighed, subjected to one of the regimes, and weighed again. Those mites that were given only 1.5 h of moist air daily showed great individual differences. Three lost weight significantly, four maintained roughly the same weight, and three gained weight significantly, showing that even under these extreme circumstances some dust mites can achieve an overall net water gain. With three and six moist hours daily, mites almost invariably regained significant net quantities of water.

Observations on *D. farinae* by Arlian et al. (1998, 1999) are also interesting in this connection. The mites were exposed to daily cycles of RH above and below CEH. Speed of development and egg production were positively correlated with the duration of the moist period. In one series of experiments at 20°C the RH during the moist period was 75% and during the dry period 0%. When only 2 or 4 h of moist air was given daily, most of the eggs (84%–92%) hatched. But the emerging offspring did not complete the life cycle or only very slowly. Many of them (44% and 53%) survived for 70 days, but not beyond the larval or nymphal stage. Nevertheless, when the surviving mites were transferred to a constant 75% RH, they completed their development to adults, demonstrating their plasticity.

Brandt and Arlian (1976) exposed *D. farinae* to constant dehydrating conditions at several temperatures in order to determine the LT-100. The LT-100 is the time needed to achieve 100% mortality. At 34°C and 50% RH it took only 6 days until all mites (150 females) were dead. At lower temperatures it takes longer. But even at 25°C all mites (140 females) were dead within 11 days. It is a pity that the LT-100 at still lower temperatures was not determined, because 11 days contrasts quite sharply with the 10-week survival of *D. pteronyssinus* that we found at 16°C.

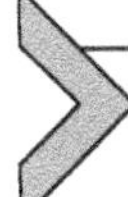

4.4 The conservation of body water, excretion of nitrogenous waste, and water loss tolerance

An important concept in biology concerns the relation between the volume of an organism and its outer surface area. Look at Fig. 4.4. The eight cubes at the right collectively have the same volume as the cube on the left. But together they have much more surface area. Smaller objects have more

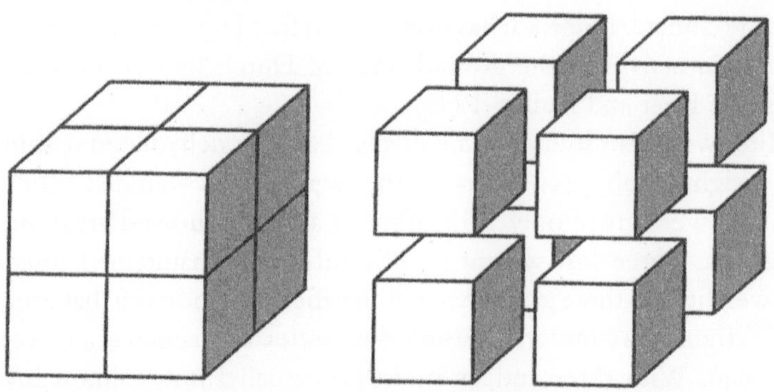

Fig. 4.4 Smaller objects have more surface area relative to their volume.

surface area relative to their volume than larger ones of the same shape. That is the reason why a big stone falls to the ground much faster than a grain of sand. It is also the reason why a mouse needs much more food, per gram of body weight, than an elephant. Also a small cup of hot water cools to room temperature more quickly than a kettle full of it. And a drop of water vaporizes much faster when it is divided into many small droplets. Evaporation, cooling, and air resistance against a moving object are all processes that depend on surface area. Such processes have a greater impact for small bodies than for larger ones. At 20°C and 60% RH a drop of water of the same size as a house dust mite will vaporize in a very short time, even in still air. Yet house dust mites thrive at a temperature of 20°C combined with a relative humidity of 60%RH. How can they?

First of all, their integument must be extremely water tight. The wax layer of the epicuticle of a terrestrial arthropod is generally recognized as a waterproofing layer. Disruption of the wax layer by abrasion or otherwise accelerates water loss. So an intact wax layer prevents the entrance and escape of water molecules. But what about fluxes of the respiratory gases O_2 and CO_2 which have a higher molecular mass than H_2O? Ellingsen (1974, 1978) reported some surprising findings about this (see Section 4.5.3).

Evaporation also depends on the humidity of the air that surrounds a mite. When mites tightly cluster together their sweating will raise the humidity in the spaces between them. Glass et al. (1998) demonstrated, for adult *D. farinae* at 0% RH, that water loss through evaporation can be substantially reduced when the mites cluster together (Figs. 4.5 and 4.6).

These authors also report another interesting but puzzling observation, namely that under optimal conditions (75% RH and 22–24°C) males but not females show a tendency to form clusters. It was suggested that pheromones are involved, but an ecological explanation is wanting. In rearing chambers quiescent protonymphs of *D. farinae*, but also other life stages, often cluster in large numbers in corners and around the edges. But how the mites behave in the domestic house dust ecosystem remains conjectural. Perhaps locally in carpets or beds the mite density can be so high that forming clumps is a feasible option for mites under desiccating conditions.

Excretion of waste metabolites may entail the loss of much water in many organisms, but not in dust mites. Part of the food that animals eat is protein. Much of the nitrogen-containing end products of metabolism is expelled as waste. Humans and other mammals convert the nitrogen-containing waste products into the nontoxic compound urea. Urea is soluble in water. In order to get rid of it we use lots of water. In contrast, mites, like birds, do not urinate. Birds convert nitrogenous waste into uric acid. Mites make guanine out of it. Uric acid and guanine are not soluble in water. These

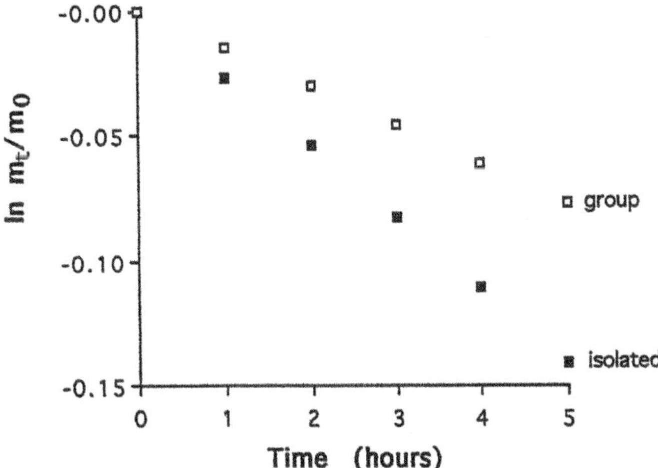

Fig. 4.5 Mean water loss at 22–24°C for single *D. farinae* males at 0% RH and for males in a group of six. The mites were starved for 12 h and predesiccated by 4%–6% of the original body mass before use. The regression slope on a semilogarithmic plot through the ratio of the water mass at time t (mt) to the original water mass (m0) over time (ln mt/m0 versus time) is the rate of water loss. Each experiment was replicated three times using 15 mites per replicate (SE ≤ 0.001). *(From Glass, E.V. Yoder, J.A., Needham, G.R., 1998 Clustering reduces water loss by adult American house dust mites* Dermatophagoides farinae. *Exp. Appl. Acarol. 22, 31–37.)*

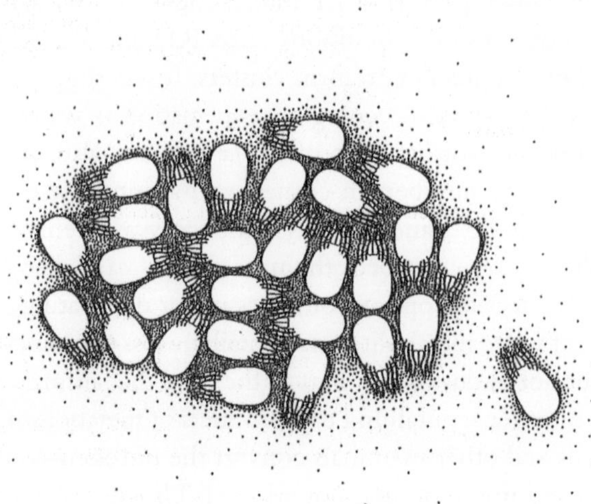

Fig. 4.6 Illustration depicting the water molecule gradient produced by clustering by adult male *D. farinae*. The greatest number of water molecules (.) was found in between clustered mites opposed to the individual. *(From Glass, E.V., Yoder, J.A., Needham, G.R., 1998. Clustering reduces water loss by adult American house dust mites Dermatophagoides farinae. Exp. Appl. Acarol. 22, 31–37 who acknowledge and thank David Dennis for the mite-clustering illustration.)*

compounds form crystals that can be excreted as part of the feces, without loss of water.

The ability to survive the loss of much body water is another adaptation that enables a mite to overcome periods with low air humidity. Among different species of terrestrial arthropods water loss tolerance varies considerably. For instance, the beetle *Carabus* spec. will die if more than 17%–22.5% of the original water content is lost. In contrast, *D. farinae* has been shown to survive a loss of 52% of its original water mass. This is one of the highest reported water loss tolerances (Arlian and Veselica, 1979). Undoubtedly, the water loss tolerance of *D. pteronyssinus* is also very high (see Fig. 4.1). Thus the hemocoel functions as a water reservoir. The striae on the integument, that are characteristic for the *Psoroptidia*, are not just superficial furrows. Cross sections reveal that these striae are true corrugations that will improve the plasticity of the integument (Fig. 4.7). These corrugations may have evolved to improve the adaptability of the volume of the hemocoel.

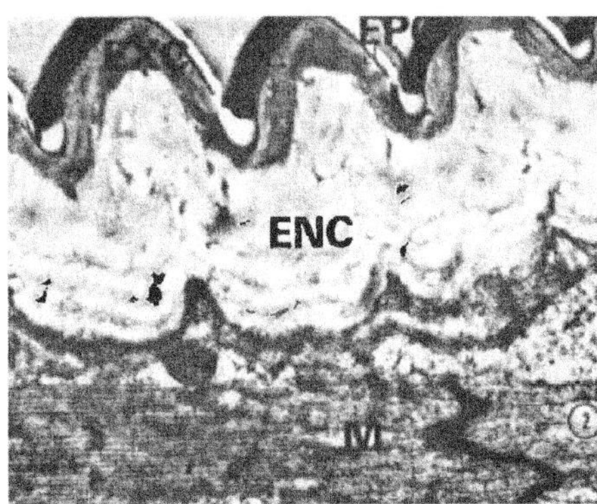

Fig. 4.7 A transmission electron micrograph of a cross section through the corrugated surface cuticle of a house dust mite, *Dermatophagoides farinae*. Three cuticular layers are obvious: the outermost epicuticle (EPC), hard exocuticle (EXC), and fibrous endocuticle (ENC), muscle fibers (M) lie beneath the cuticle. 10,000×. *(From: Brody, A.R., 1971. Microscopic anatomy of the house dust mite, Proc. North Central Branch E.S.A. 26, 64–65.)*

4.4.1 The mechanism of water vapor uptake

Active uptake of water molecules from the unsaturated ambient air into the mite's body is an energy demanding process. It involves transport of molecules against a gradient (Fig. 4.8.). That speaks for the involvement of the supracoxal glands, because the cells of these glands are rich in mitochondria, the power stations of cells. Energy-carrying ATP molecules are manufactured in these organelles.

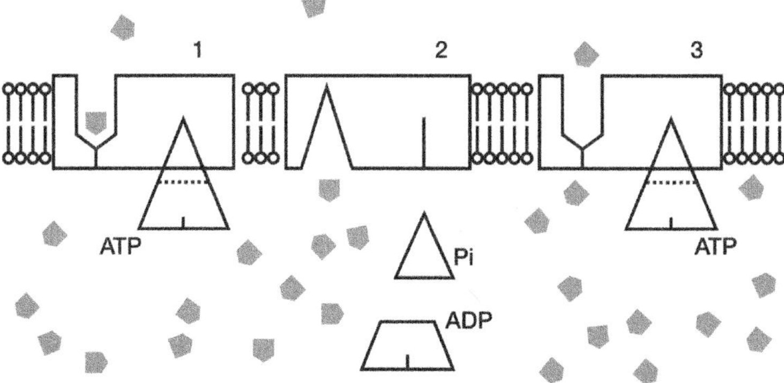

Fig. 4.8 Active transport of a solute (gray symbol) across a membrane, mediated by a transport enzyme. Three stages of the transport are shown. A solute molecule is captured (1) and forcefully replaced to the other side of the membrane at the expense of an ATP molecule (2). Thereafter the transport enzyme is restored (3). *(Artwork by the author.)*

Arnold Brody et al. (1976) studied the supracoxal glands of *D. farinae* using light microscopy as well as electron microscopy. Each one of the supracoxal glands is made up of eight cells. Brody distinguished four cell types, A, B, C, and D. Cells of type A have fimbriae that extend into the hemocoel. These fimbriae constitute an enlargement of the cell membrane which may indicate massive transport of some compound or compounds across the cell membrane between the hemocoel and the interior of the gland cell. Many mitochondria are associated with this membrane, indicating energy-requiring processes such as transport of molecules against a gradient. At the other end of the gland cell, minute cuticle-lined ducts can be discerned inside the cell. According to Brody's interpretation, these ducts communicate with similar ducts in the other gland cells and with the outside world. Cells of type C have an extensive cytoplasmic reticulum, indicating synthesis of proteins. The various gland products may be mixed before they are discharged and exposed to the outside air. In an earlier study (Brody et al., 1972) it was observed that "at the level where the hypognathosoma articulates with the idiosoma, a channel partially closed from the outside by a cuticular flap crosses over the dorsal margin of the palp and delivers glandular secretions into the prebuccal cavity." Thus these glandular secretions are exposed to the air so they can absorb water molecules. Then the mixture is discharged into the prebuccal cavity where it can be ingested.

The observed dynamics of water loss of mites that have been transferred from high to low humidity (see Figs. 4.9 and 4.10) can be explained by postulating that the water content of a mite is divided between two "compartments." One of the compartments dries up very fast and the other one dries up slowly. The fast drying compartment allegedly concerns the water that is under way from the supracoxal gland to the prebuccal cavity (Arlian, 1975). Supposedly, water vapor uptake is accomplished by exposing a hygroscopic material, produced by the supracoxal glands, to the ambient air. This hygroscopic material is ingested together with the absorbed water molecules. Somehow it must flow from the supracoxal glands to the buccal cavity where it can be ingested.

Arlian and Wharton (1974) loaded *D. farinae* mites with tritium-labeled (radioactive) water (HTO) and allowed them to equilibrate at 75%RH.

Body mass (μg) and tritium content (counts per minute) were monitored after the mites were transferred to either of seven test humidities. The water content was determined by subtracting the dry weights. Three of the test humidities were below the CEH and five were above the CEH. The data were presented in a table. I plotted them in a graph (Figs. 4.9 and 4.10). Not every detail of the procedure followed by the authors is clear. Individual mites were weighed and measured sequentially, but the number of replicate observations is not mentioned. Standard errors are substantial in comparison with the averages (3%–23% for water mass and 4%–31% for tritium content).

In Fig. 4.9A, the decline of the water content of mites exposed to humidities below the CEH is shown. During the first period the decline is stronger than in later periods. This can be interpreted as the rapid evaporation of water in the hygroscopic fluid that is exposed to the air on its way from the supracoxal glands to the mouth, i.e. the water of the pump mechanism. After the water in the pump has disappeared, it is not unreasonable to expect that, in each time unit, escaping water molecules constitute a fixed proportion of the water molecules present at the start of it. If that is so, the logarithms of the water mass, when plotted against time, will be on a straight line. Therefor the logarithms, instead of the true values, are plotted in the figures. Deviations from a best fitting straight line may be caused by unavoidable experimental inaccuracies.

Fig. 4.9B shows that, above the CEH, mites are perfectly able to compensate the water losses by active uptake of water vapor. This absorbed water has no tritium labels. Fig. 4.10A and B shows how fast the tritium-labeled water molecules escape from the mites. It is obvious that at humidities above the CEH (Fig. 4.10B) labeled molecules escape at a faster rate.

Wharton and Furumizo (1977) provided further evidence. They kept *D. farinae* mites in dry air and observed an extensive "stalactite-like" plug in the region of the orifice of the supracoxal glands of these mites. These plugs deliquesced when exposed to high humidities. By "electron probe elemental analysis" for elements heavier than fluorine, the plugs were found to contain concentrations of K and Cl.

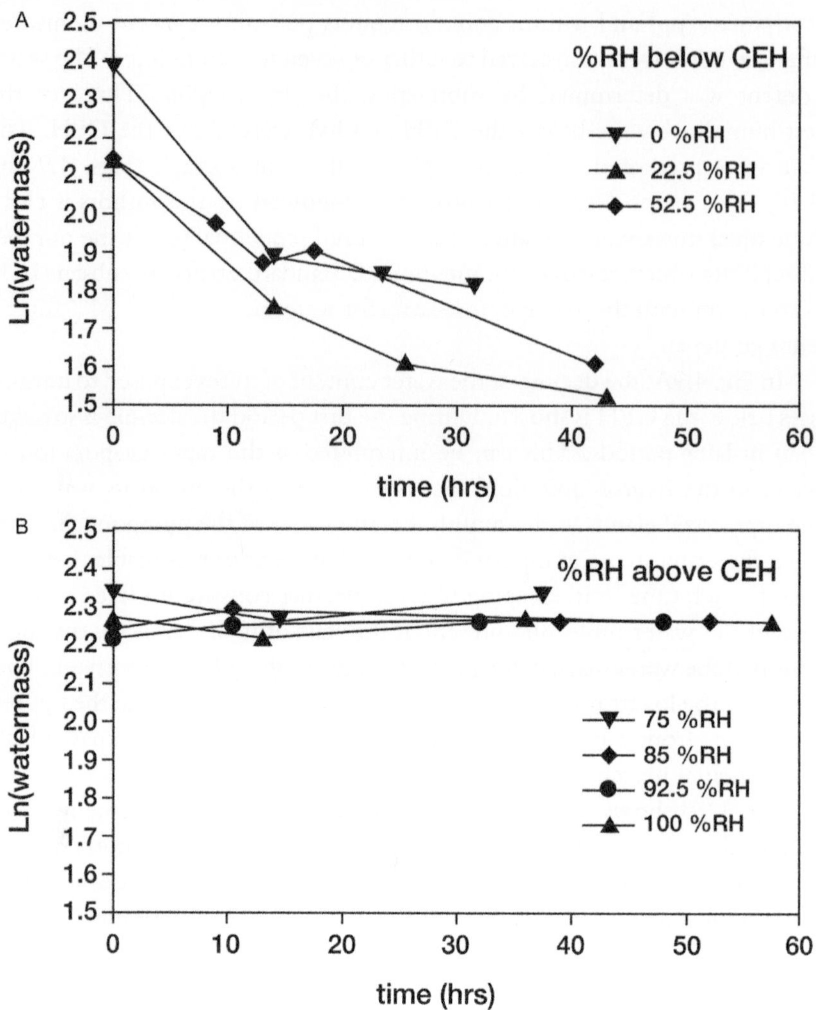

Fig. 4.9 Evolutions of total body water mass of *D. farinae* females, which were transferred from 75% RH to either of seven test humidities. *(Based on data from Arlian, L. G., Wharton, G.W., 1974. Kinetics of active and passive components of water exchange between the air and a mite, Dermatophagoides farinae. J. Insect Physiol. 20, 1063–1077.)*

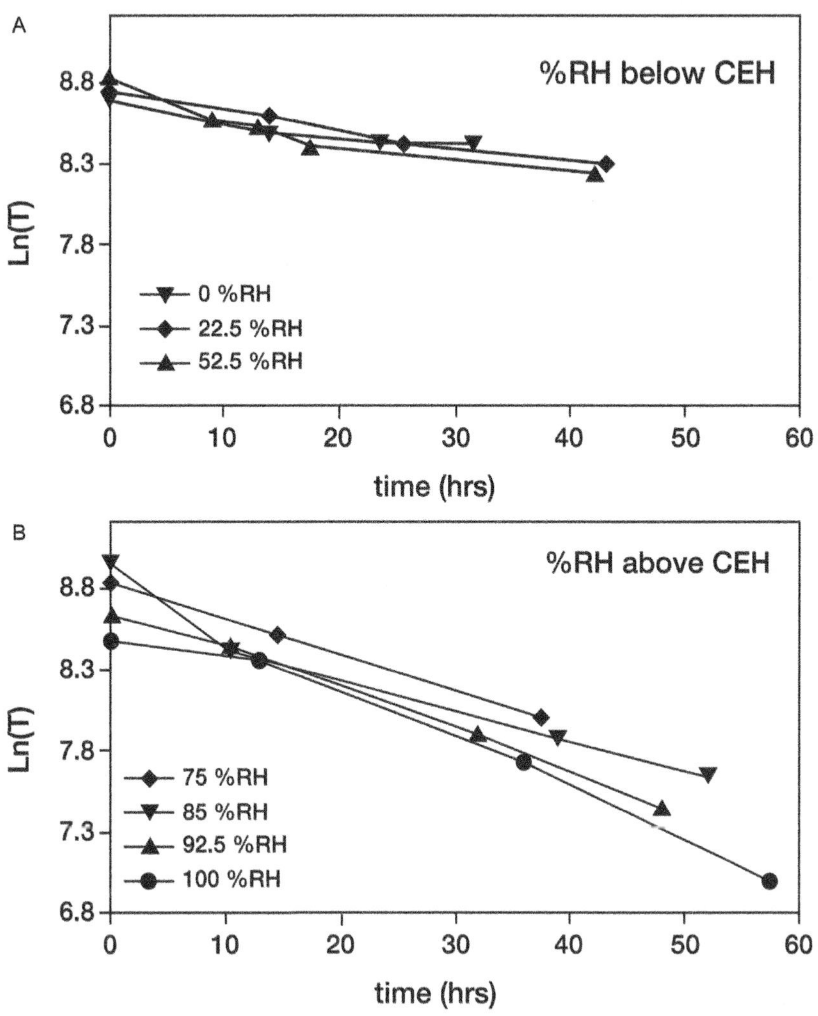

Fig. 4.10 Evolutions of tritium-labeled body water (counts per minute) of *D. farinae* females, which were transferred from 75% RH to either of seven test humidities. *(Based on data from Arlian, L.G., Wharton, G.W., 1974. Kinetics of active and passive components of water exchange between the air and a mite,* Dermatophagoides farinae. *J. Insect Physiol. 20, 1063–1077.)*

4.4.2 Identity of the hygroscopic substance

We don't know the identity of the hygroscopic substance that serves as a vehicle for transport of water from the atmosphere into the mite. Let us not forget that its existence is hypothetical. If it exists it must be hygroscopic enough to extract water vapor from the air at a relative humidity just a little above 58% at 25°C (the CEH of *D. farinae*) and remain liquid. Certainly an interesting compound or mixture of compounds. It must be possible to obtain an amount of it in pure solution. Rearing mites in large quantities is not difficult. It is also possible to separate these mites from the culture medium. Put some culture on a glass plate and drive the mites out of it using a heat source and harvest gram quantities of pure and clean mites with a small brush. Probably the compounds that we are interested in are soluble in water and can be washed off. Subsequent steps are up to modern chemistry.

4.5 Two strategies for dealing with drought spells: *D. pteronyssinus* vs *D. farinae*

Before I turned to the study of dust mites I worked on spider mites. Spider mites infest plants and feed on the leaves. The spider mite *Tetranychus urticae* survives the winter in a state that is called 'diapause.' The induction of diapause is a deterministic process. It is determined by day length, i.e. the number of hours of daylight per day. Spider mites that grow up with less than a certain length of daylight enter into diapause. They stop feeding and don't lay eggs. They become positively geotactic and negatively phototactic and go for shelter. Intense pigmentation makes them also look totally different from the active mites, making the transition a dramatic scene for a human observer. In nature the decreasing day length is a sure sign that winter is approaching. And in winter the plants that spider mites live on die or shed their leaves. It is a period when spider mites cannot feed and cannot reproduce. All they can do is try to survive. During diapause their physiology and behavior is changed in a way that gives them a chance to survive. When winter is approaching, all spider mites enter into diapause—100% of them. And that makes sense. An odd mite that would fail to enter into diapause has 0% chance to survive. Winter is a hard time for spider mites but, fortunately for them, the advent of winter is predictable. Day length for them is like a guiding angel.

Dust mites also sometimes have to cope with periods of unfavorable conditions. But the occurrence of these periods is not predictable. A great danger that dust mites have to cope with is the incidental occurrence of a lengthy period with low air humidity. If such a period lasts long enough, a population of *D. pteronyssinus* will die out completely. A population of the other species, *D. farinae,* however, would have a chance to survive. The chance to survive depends on a peculiarity in their development. A developing mite passes through three immature mobile stages. Normally a quiescent (immobile) protonymph gives rise to a (mobile) tritonymph within a few days. But in *D. farinae* a varying proportion of the quiescent protonymphs remains in the quiescent stage for many months. And in that condition they are much more resistant to desiccation than the mobile forms.

In contrast with the induction of diapause in spider mites, the prolongation of the quiescent stage in *D. farinae* is a stochastic phenomenon, not a deterministic one. Not all quiescent protonymphs remain quiescent for a long time. Only a varying proportion does.

When after a period with desiccating circumstances the humidity of the air rises again to a level above the CEH, all surviving *D. pteronyssinus* mites will immediately start to feed and reproduce. That gives them an advantage over competing *D. farinae.* The *D. farinae* mites that have entered into prolonged quiescence must wake up before they can start feeding and reproducing. That will take time. Even when humidity conditions would allow them to be active, feed, and lay eggs, part of the population remains inactive as an insurance for the possible occurrence of disastrous drought. So there is a disadvantage attached to the propensity of *D. farinae* to enter into prolonged quiescence. But, occasionally, it allows them to escape total destruction. In climates where long lasting uninterrupted drought spell occur frequently *D. farinae* must be in an advantageous position toward competing *D. pteronyssinus.*

Prolongation of quiescence in *D. farinae* occurs mainly in the protonymphal stage. In the sequence of life stages, the prolonged quiescence occurs at the point where in many other astigmatic mites an extra molt can be intercalated, giving rise to a hypopus (see Section 7). Maybe just a chance coincidence. Maybe not. Also in the tritonymphal stage of *D. farinae* prolongation of quiescence has been observed, but less often than in the protonymphs. Some authors refer to it as "resting," but also the term "diapause" has been used for it. The occurrence of prolonged quiescence in *D. farinae* has been observed many times in laboratory cultures maintained for the production of allergens needed for diagnostic or therapeutic purposes. The phenomenon is sometimes experienced as a nuisance because it interferes with a desired, unrestrained population growth. Observation of the phenomenon in more natural situations is not so easy. Changes in the proportional abundance of the various life stages during the year can sometimes be interpreted as evidence of the occurrence of prolonged quiescence. The following citations may illustrate this:

> From October till May the D. farinae *population is primarily composed of protonymphs. In the period of maximum population density (May–October) the presence of adults is predominant (Dusbabek, 1975).*
>
> *The population structure of D.f. changed by season; relative increase of nymphs in winter populations was characteristic, suggesting appearance of some desiccation-tolerant nymphs. ... No marked difference was seen in the population structure of D.p. between summer and winter ... (Suto et al., 1991)*

Arlian et al. (1983) monitored the abundance and population structure of dust mites in Ohio (USA). Roughly once every month a dust sample was taken from couches, floors, and mattresses of five homes during 1½ years. Most of the time adult males and females were the most abundant stages, but in April and May the numbers of tritonymphs increased dramatically, to constituting more than 75% of the live mites collected. Most likely, these were *D. farinae* tritonymphs that survived the winter as quiescent protonymphs. (Unfortunately *D. farinae* and *D. pteronyssinus* were not distinguished in this study.)

Matsumoto et al. (1986) studied the life cycle of *D. pteronyssinus* and of *D. farinae* at 25°C at three different humidities: 61%, 76%, and 86%RH. The duration of each of the immature stages, egg, larva, protonymph, and tritonymph was typically a little longer than 1 week. Protonymphs of *D. farinae* that took less than 10 days to molt into tritonymphs were classified as "regular." Very few protonymphs molted after 11–20 days, but quite a few molted after 20–40 days. These were classified as "prolonged." At 61% RH, 18 out of 22 were prolonged. At 76% RH, 8 out of 19 were prolonged. At 86% RH, 11 out of 17 were prolonged. This phenomenon, a prolonged protonymphal stage, was not seen in *D. pteronyssinus*. Not only the quiescent protonymph of *D. farinae,* but also the quiescent tritonymph stage could be prolonged. However, that happened far less frequently.

Ellingsen (1978) collected quiescent protonymphs for her physiological studies and reported the following observations: "Quiescent forms of the protonymph were found on the top of the culture vials. They showed tigmotactic behavior before becoming quiescent and were thus found on the rim of the vials where the cover snaps on tightly, or in the corners of the covering paper. These mites had produced a sticky secretion that anchored them to the surface."

4.5.1 Induction of (prolonged) quiescence of *D. farinae*

Although the phenomenon is a stochastic one, external factors do influence the probability of entering into prolonged quiescence. Ellingsen (1974) states that "In the life-cycle of the American house-dust mite, quiescent nymphs are formed facultatively; 89% of them are protonymphs and 11% are tritonymphs. They occur in cultures under optimum conditions of temperature and humidity, and their formation may be stimulated by a crowding factor. The quiescent protonymphal period may last for 60 days or more ...".

Reka et al. (1992) found evidence for the existence of an aggregation pheromone in *D. farinae* and surmise that it also plays a role in the induction of resting in protonymphs.

A series of Japanese papers with abstracts and figure captions in English, published during the years 1990 through 1995, present valuable information concerning the factors that influence induction and termination of prolonged quiescence. The field observations and the experiments were done in and around the city of Nagoya, Japan. Here *D. pteronyssinus* and *D. farinae* are about equally common. Sakaki et al. (1990) collected mites from the rims of crowded cultures and placed them at 25°C and 75%RH without food. Within 1 month about 30% of them, mostly protonymphs, entered into the quiescent state. Conditions in the cultures may have triggered their inclination to become quiescent. Half of them remained quiescent for 5 months at 25°C and 75% RH and a further 5 months at room temperature. Suto and Sakaki (1990) presented experimental evidence that the %RH influences the proportion of nymphs entering prolonged quiescence (Fig. 4.11). At 75% RH and at 85%RH many mites become quiescent within about 3 weeks. At 33% and 43%RH, they don't. These humidities are below the CEH, so the mites must have been partially dehydrated. Perhaps the mites enter into quiescence only if they are fully hydrated. At 100%RH they did not become quiescent either. That is more difficult to explain. Mold growth is often a problem when cultures are kept at 100%RH. The authors themselves do not say anything about it, at least not in the English part of their paper.

The same authors confined groups of 10, 100, 280, or 600 nymphs in a "shallow hole on slide glasses." With increasing inoculum size, the mites were more inclined to become quiescent.

In summary, there is evidence that crowding stimulates the propensity to enter into prolonged quiescence, but only in fully hydrated mites.

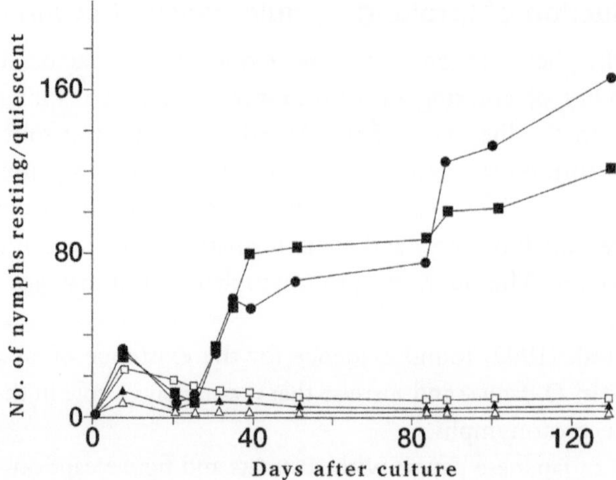

Fig. 4.11 Effect of RH on the occurrence of resting and/or quiescent nymphs of *D. farinae*. Groups of 100 mites were separately kept in a shallow hole (2.5 cm in diameter and 1.5 mm deep) on a slide glass with 50-mg culture medium and incubated at 25°C. RH: 85% (●), 75% (■), 43% (▲), 33% (△), and 100% (□). *(From Suto, C., Sakaki, J., 1990. Studies on the factors influencing the induction, persistance and termination of prolonged quiscence (diapause) in house dust mite,* Dermatophagoides farinae. *Jpn. J. Sanit. Zool. 41, 375–381.)*

4.5.2 Termination of (prolonged) quiescence of *D. farinae*

The seasonal increase of tritonymphs in house dust (Section 4.5) indicates that the termination of prolonged quiescence is somehow influenced by physical conditions. The factors that influence termination were also studied by Suto and Sakaki (1990) (see: Figs. 4.12 and 4.13). They write: "Termination of quiescence indicated by moulting to tritonymphs was accelerated when the quiescent nymphs were disturbed by separation from the substrates to which they had clung. The quiescent period of undisturbed nymphs was estimated to be about seven months at a constant condition of 25°C and 75% RH." Also Sakaki and Suto (1991) mention that separation from the substrate triggered molting in a large part but not in all the quiescent nymphs, which illustrates the stochastic nature of these processes. Separation from the substrate, presumably with a brush, is of course a very artificial kind of disturbance. But it may resemble the disturbances experienced by a pyroglyphid mite living in the nest of a bird or a mouse, as the forebears of house dust mites presumably did. Contact with water was also shown to stimulate termination of quiescence. An ecological explanation for that is less obvious.

For subsequent experiments, quiescent mites were detached from the substrate and transferred to a special confinement for exposure to various

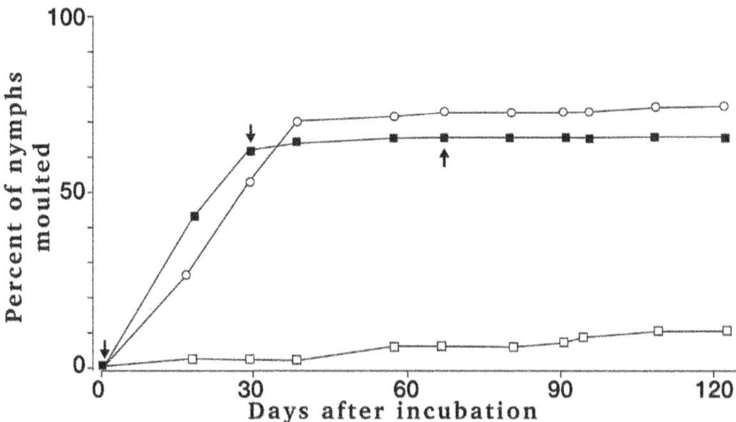

Fig. 4.12 **Termination of quiescence.** Effect of disturbance on molting of the nymphs which had been kept for 4 weeks at 25°C. 75%RH. The quiescent nymphs were disturbed by separation from the substrate to which they had clung or by being transferred. They were incubated in the shallow hole (1.5 cm in diameter and 0.5 mm deep) on slide glasses without food. □, adhering to the substrate; ○, disturbed once; ■, disturbed three times (arrows). *(From Suto, C., Sakaki, J., 1990. Studies on the factors influencing the induction, persistance and termination of prolonged quiscence (diapause) in house dust mite, Dermatophagoides farinae. Jpn. J. Sanit. Zool. 41, 375–381.)*

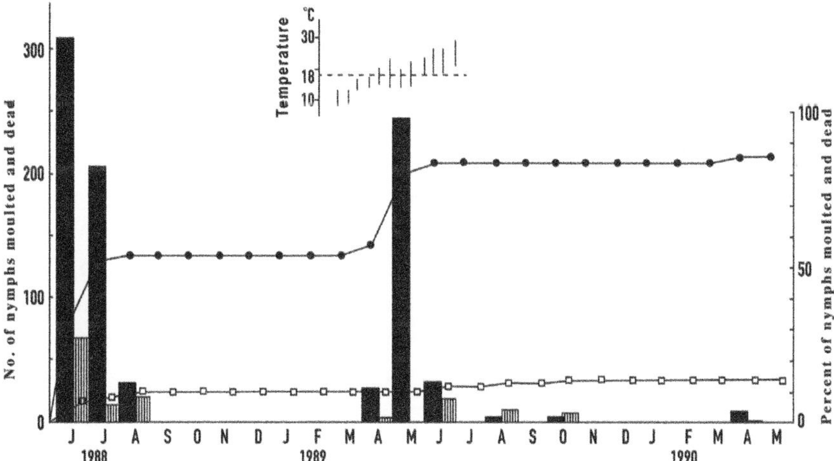

Fig. 4.13 Number and cumulative percentage of molt and death in resting/quiescent nymphs of *D. farinae* (Group A, 1045 nymphs in total), and maximum and minimum temperatures at the termination of overwintering. Mites dispersing from culture bottles were collected in late May 1988. They were incubated at 25°C and 75%RH, without food until November, thenceforth at room temperature. Overwintered nymphs, 378 in 1988; 18 in 1989. ■, No. nymphs molted; ▨, no. nymphs dead; ●, percent of nymphs molted; □, percent of nymphs dead. *(From Sakaki, J., Suto, C., 1991. The termination of prolonged quiscence (diapause) in the house dust mite, Dermatophagoides farinae at room temperature. Jpn. J. San. Zool. 42, 93–97.)*

experimental circumstances. It could be shown that the age of the quiescent nymphs had a marked influence. Nymphs of 6, 8, 20, and 32 weeks old showed an increasing tendency to end quiescence. Cooling for 1 week to 5°C enhanced termination of quiescence, but cooling to 10°C or 18°C did not or not so strongly. Graphs in the paper by Sakaki et al. (1990) also show that the occurrence of molting is strongly influenced by both temperature and %RH. Molting of nymphs aged 22 weeks or more was suppressed by incubating them at temperatures below 25°C or RH below 55%.

Sakaki and Suto (1991) isolated mites that became quiescent at different times of the year ranging from May 1988 to September 1989. These were placed in a "house room," confined in a plastic box over a saturated NaCl solution. "Most of the nymphs in each group overwintered and terminated quiescence almost synchronously during the period from April to June (mainly May) of the following year, irrespective of the time when they entered into quiescence. A few nymphs continued quiescence for over one year and most of them overwintered again and also moulted in April" (Fig. 4.13).

The possible influence of day length was not investigated.

4.5.3 Physiology of quiescent *D. farinae* nymphs

Ellingsen (1975) used a radioactive label to study the loss of water molecules from quiescent as well as active protonymphs of *D. farinae*. Mites loaded with tritiated water (HTO) were placed in an atmosphere without tritium, and the loss of tritium from the mite bodies was monitored (Fig. 4.14.). Quiescent protonymphs initially lost water during about 16 days. This initial loss of water is believed to be caused by the injuries inflicted when the quiescent nymphs were detached from the substrate. After 16 days it seems that these quiescent mites were practically impermeable for water molecules. According to Ellingsen, quiescent protonymphs have a thicker cuticle than active mites. This may indicate that the injured quiescent mites repaired their cuticle by secretion of more wax. Active protonymphs under the same circumstances lose water continuously and do not become completely impermeable.

If it is possible to make the cuticle completely impermeable for water, you may ask, why would dust mites not have an impermeable integument all the time? Why wouldn't they go on with their usual activities like searching for food, eating, and laying eggs, while the cuticle is impermeable? There may be more than one reason. Probably an important reason is that

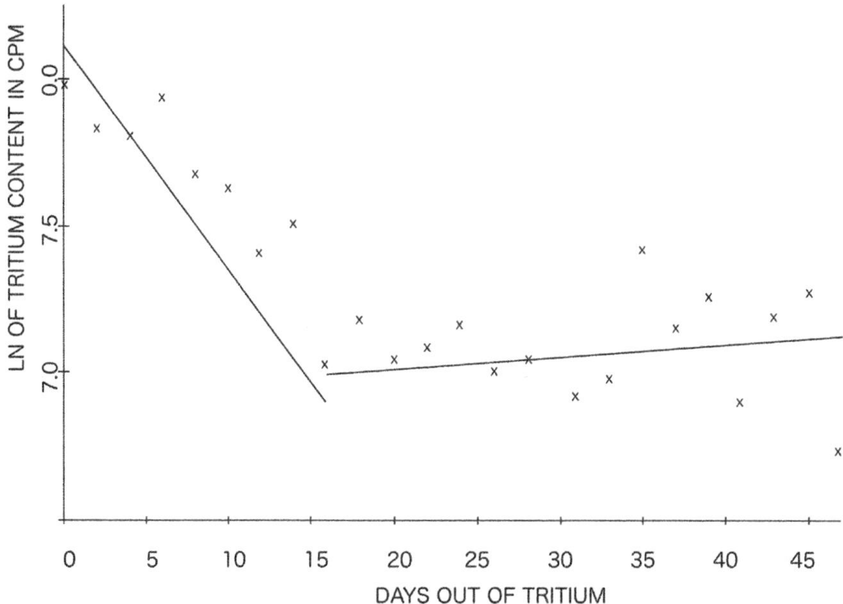

Fig. 4.14 Loss of tritium-labeled water from quiescent protonymphs of *D. farinae*. After about 16 days the slope of the regression line is slightly going up, but the slope is not significantly different from zero, indicating that these quiescent mites were practically impermeable for water molecules. *(From Ellingsen, I.J., 1975. Permeability to water in different adaptive phases of the same instar in the American house dust mite. Acarologia 17, 734–744.)*

such activities require oxygen and entail the production of CO_2. Impermeability for water may be inconsistent with the uptake of oxygen and the elimination of CO_2. Quiescent protonymphs likely have a lower metabolism than active mites. But, although quiescent protonymphs seem to be impermeable for water, they do take up oxygen, albeit at a much lower rate than active mites do. Ellingsen (1978) found that undisturbed, quiescent protonymphs of *D. farinae* consumed O_2 at a rate that was 28.5 times less than the O_2 consumption of active protonymphs.

4.5.4 The balance between *D. pteronyssinus* and *D. farinae* infestations

In geographic areas with a relatively dry climate or with very cold and dry winters, extended periods of drought presumably occur relatively often. *D. farinae* must be in an advantaged position there and may be expected to be more numerous than in other areas.

Arlian et al. (1992) examined the house dust acarofauna of 251 homes in 8 widely spread locations in the United States. Most of these homes (205) had both *D. farinae* and *D. pteronyssinus*, but 22 of them (8.8%) were inhabited by only *D. pteronyssinus* and 24 (9.6%) of the houses had only *D. farinae*. Perhaps in these last houses, where *D. farinae* was the only *Dermatophagoides* species, long, uninterrupted periods with low air humidity (RH < CEH) occurred regularly, so that *D. pteronyssinus* could not survive. Unfortunately, records of RH and temperature in the mite's habitat in these houses are not available. A clear relationship with the local climate cannot be seen in the data set provided by Arlian et al. Quite a lot of houses (251) were implicated in this study, but that number was nevertheless insufficient to demonstrate a connection between the geographic position and the composition of the mite faunas.

Data concerning the house dust fauna in many other locations all over the world can be found in the literature. In contrast with the data provided by Arlian et al., these do not give information about individual houses. The mite numbers of several houses in a location are pooled and presented as mites per gram of dust or as total numbers of mites or as mites per square centimeter, etc. Nevertheless, the relative abundance of *D. farinae* compared with *D. pteronyssinus* can be calculated. Of the 36 data sets that could be used for this comparison, 19 were from a study done in a location at the sea coast. In the 17 inland locations, *D. farinae* was found a little more often than at the coast, but the difference is not significant. If only the 17 inland locations are considered, the Spearman rank correlation statistic can be applied to see if *D. farinae* is more common in locations further away from the coast. And indeed, further away from the coast *D. farinae* is found a little more often. But again the trend is by no means statistically significant.

In five locations no *D. farinae* were found at all. These five locations have either a maritime or Mediterranean climate, which is in line with our expectation. There are also two locations where only *D. farinae* is found and no *D. pteronyssinus* at all. The climate in these two locations is classified as "continental," which again is in agreement with our expectation.

In conclusion, if the climate of a geographic location has any influence at all on the numerical balance between the two *Dermatophagoides* species, than this influence is very small. The data concerning the differences between different climates are by no means sufficient to reach statistical significance.

4.5.5 Other differences between *D. pteronyssinus* and *D. farinae*

Kuehr et al. (1994) found that home characteristics that are associated with elevated concentrations of *D. farinae* allergens are not the same as those associated with elevated concentrations of *D. pteronyssinus* allergens. Indeed *D. pteronyssinus* seems to be more sensitive for characteristics that influence air humidity (category A, see Section 5.1). The ability of *D. farinae* to form long-lived, drought-resistant protonymphs may well be the most important ecological difference with *D. pteronyssinus*. What other differences distinguish the two species? It has been noticed by me and others that *D. pteronyssinus* is far more reluctant to walk on glass surfaces than *D. farinae*. But only a limited number of strains of both species have been kept in the laboratory. Possibly strains of the same species can also be different concerning this aspect. Also differences in food requirements between the two species have been reported (Section 9.1). Hart and Le Merdy (1987) found that the most favorable single food was different for the two species. Molva et al. (2019) also found that an optimal diet for *D. pteronyssinus* is not optimal for *D. farinae* and vice versa.

Then there are morphological differences. *D. pteronyssinus* is smaller. Its body weight is roughly half that of *D. farinae* (Arlian, 1977). However, Solarz et al. (2016) found that males of *D. farinae* show great size variability. The size of females may also be variable. Solarz et al. studied the variability of males of *D. farinae* and could distinguish three heteromorphic forms, differing in several characters. The width of the conspicuous femur I (Fig. 4.17.) was also very variable. This phenomenon, the occurrence of heteromorphic males, is called "andropolymorphism." It has been recorded in five species of the genus *Dermatophagoides* (*D. farinae*, *D. anisopoda*, *D. neotropicalis*, *D. sclerovestubilatus*, and *D. simplex*). It is remarkable therefore that it has not been recorded for *D. pteronyssinus*, whereas thousands of specimens of this species have been examined by acarologists.

Distinguishing species by morphological characters is important for the performance of ecological studies. In all stages except the egg stage, the two species can be distinguished by examining the position of the setae sci and sce (Fig. 4.15.). These are trifling differences but they can be seen without extra magnification. I always looked at this character in the first place, but when the mite was an adult female I habitually checked the diagnosis by examining the receptaculum seminis. It was always right and that strengthened my confidence. Adults of the two species can be distinguished clearly (Fig. 4.16).

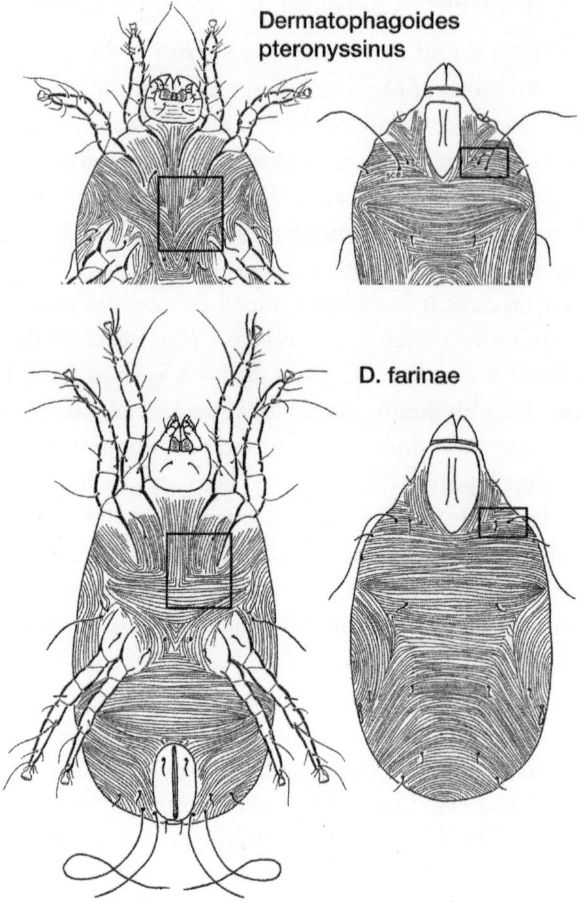

Fig. 4.15 Two characters distinguishing *Dermatophagoides pteronyssinus* from *D. farinae*: (1) On the ventral side (left) the pattern of striations around coxa I and II is acute, V-shaped in *D. pteronyssinus*, and right-angled in *D. farinae*. (2) On the dorsal side (right) the position of the inner supracoxal seta (sc i) relative to the external supracoxal seta (sc e) is different.

For species that commonly occupy the same habitat, as *D. pteronyssinus* and *D. farinae* do, it is of vital importance to have a prezygotic mating barrier. Imagine there would be no prezygotic mating barrier. Then males and females of different species would engage in a copulation when they meet. Zygotes (fertilized eggs) would be formed by the fusion of egg cells from one species and sperm from the other one. The resulting hybrid zygotes would either be nonviable or give rise to hybrid individuals that are sterile or poorly adapted to their environment. Thus it is advantageous for both species to prevent the formation of such hybrid zygotes by putting up a mating barrier.

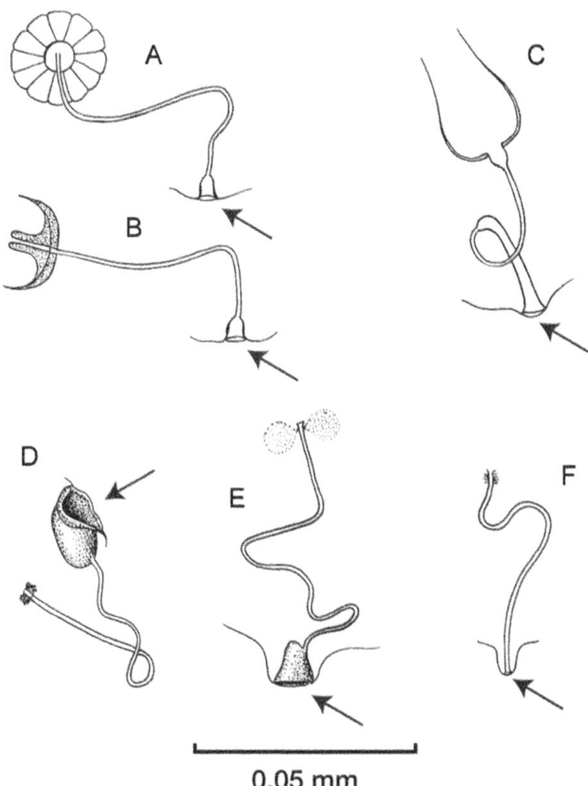

Fig. 4.16 Sclerotized parts of the female genitals of six mite species: (A) and (B) *Dermatophagoides pteronyssinus* in two different orientations, (C) *D. evansi*, (D) *D. farinae*, (E) *D. rwandae*, (F) *Hirstia passericola*. Arrows indicate the Bursa copulatrix. *(From Fain, A., Guerin, B., Hart, B.J., 1990. Mites and Allergic Disease. In: B. Guerin (Ed.), Allerbio, Varenne en Argonne, France, 190 pp.)*

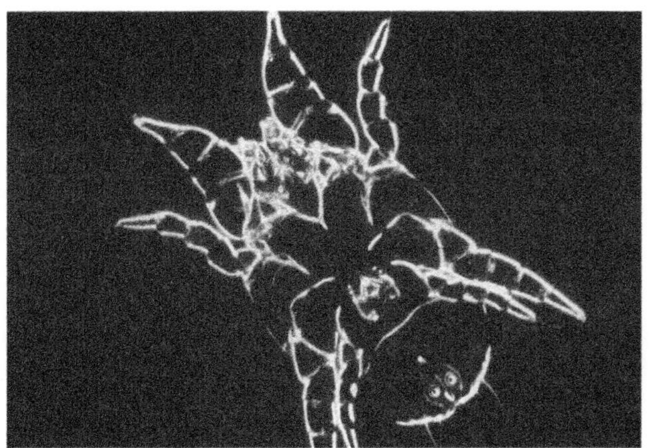

Fig. 4.17 Male *D. farinae*. Note the size of the front legs. *(Photo: Kees Kuller.)*

The best mating barrier is a behavioral barrier. Then males are simply not interested in females of the other species and vice versa. No time is wasted on fruitless attempts to copulate. If no such barrier exists then males and females of different species will engage in copulation attempts. But even then sperm transfer may be prevented. In order to mate successfully the genitals of the mating partners must be a perfect fit. If genitals don't fit, sperm transfer is prevented. A barrier of that kind is referred to as a lock-and-key mechanism; the key must have the right shape to fit into the lock. Indeed in closely related, similar looking arthropods, differences in the morphology of the genitals often provide the clearest visible distinction between two species. That is also true for *D. pteronyssinus* and *D. farinae*. Fig. 4.16. shows the sclerotized parts of the female genital organs of six mite species that otherwise look very similar. It confirms the line of reasoning given before, because, as you can see, the genitals are quite distinct. However, one thing has always bothered me. The eye-catching differences between the genitals of the different species are not in the region of the bursa copulatrix, where male and female copulatory organs make their first contact. The most striking differences are seen in the receptaculum seminis, deep inside the female body, where the penetrating penis can hardly reach. Perhaps the resemblance with a lock and key must not be taken too literally. What we see of the genitals are only the sclerotized parts, hard and inflexible, like a real lock and key. But the differences in the sclerotized parts may well reflect more important differences in the soft tissues, muscles, and nerves that surround it—the soft body parts that are indispensable for sperm transfer. The sperm of *Dermatophagoides* is nonmotile (Walzl, 1992). It must be actively injected by the male and taken up by the female. That, most likely, requires a delicate cooperation between the partners.

References

Arlian, L.G., 1975. Dehydration and survival of the European house dust mite, *Dermatophagoides pteronyssinus*. J. Med. Entomol. 12, 437–442.

Arlian, L.G., 1977. Humidity as a factor regulating feeding and water balance of the house dust mites *Dermatophagoides farinae* and *D. pteronyssinus*. J. Med. Entomol. 14, 484–488.

Arlian, L.G., Veselica, M.M., 1979. Water balance in insects and mites (review). Comp. Biochem. Physiol. 64, 191–200.

Arlian, L.G., Veselica, M.M., 1981a. Effect of temperature on the equilibrium body water mass in the mite *Dermatophagoides farinae*. Physiol. Zool. 54, 393–399.

Arlian, L.G., Veselica, M.M., 1981b. Reevaluation of the humidity requirements of the house dust mite *Dermatophagoides farinae*. Med. Entomol. 18, 351–352.

Arlian, L.G., Wharton, G.W., 1974. Kinetics of active and passive commponents of water exchange between the air and a mite, *Dermatophagoides farinae*. J. Insect Physiol. 20, 1063–1077.

Arlian, L.G., Woodford, P.J., Bernstein, I.L., Gallagher, J.S., 1983. Seasonal population structure of house dust mites, *Dermatophagoides* spp. (Acari: Pyroglyphidae). J. Med. Entomol. 20, 99–102.

Arlian, L.G., Bernstein, D., Bernstein, I.L., Friedman, S., Grant, A., Lieberman, P., Lopez, M., Mezger, J., Platts-Mills, T., Schatz, M., Spector, S., Wasserman, S.I., Zeiger, R.S., 1992. Prevalence of dust mites in the homes of people with asthma living in eight different geographic areas of the U.S. J. Allergy Clin. Immunol. 90, 292–300.

Arlian, L.G., Neal, J.S., Bacon, S.W., 1998. Survival, fecundity, and development of *Dermatophagoides farinae* (Acari: Pyroglyphidae) at fluctuating relative humidity. J. Med. Entomol. 35, 962–966.

Arlian, L.G., Neal, J.S., Vyszenski-Moher, D.L., 1999. Fluctuating hydrating and dehydrating relative humidities effects on the life cycle of *Dermatophagoides farinae*. Med. Entomol. 36, 457–461.

Brandt, R.L., Arlian, L.G., 1976. Mortality of house dust mites, *Dermatophagoides farinae* and *D. pteronyssinus*, exposed to dehydrating conditions or selected pesticides. J. Med. Entomol. 13, 327–331.

Brody, A.R., McGrath, J.C., Wharton, G.W., 1972. *Dermatophagoides farinae*, the digestive system. N. Y. Entomol. Soc. 80, 152–177.

Brody, A.R., McGrath, J.C., Wharton, G.W., 1976. *Dermatophagoides farinae*: supracoxal glands. N. Y. Entomol. Soc. 84, 34–37.

de Boer, R., 2000. Rates of water vapour uptake and loss by the house dust mite, *Dermatophagoides pteronyssinus* (Acari: Pyroglyphidae). Proc. Exp. Appl. Entomol. N.E. V. Amsterdam 11, 73–76.

de Boer, R., Kuller, K., 1995. Winter survival of house dust mites (*Dermatophagoides* spp.) on the ground floor of Dutch houses. Proc. Exp. Appl. Entomol. N.E.V. Amsterdam 6, 47–52.

de Boer, R., Kuller, K., Kahl, O., 1998. Water balance of the house dust mite, *Dermatophagoides pteronyssinus* (Acari, Pyroglyphidae), maintained by brief daily spells of elevated air humidity. J. Med. Entomol. 35, 905–910.

Dusbabek, F., 1975. Population structure and dynamics of the house dust mite *Dermatophagoides farinae* in Czechoslovakia. Folia Parasitol. 22, 219–231.

Ellingsen, I.J., 1974. Comparison of Active and Quiescent Protonymphs of the American House-Dust Mite. Dissert. Abstracts:, pp. 3740–3741B.

Ellingsen, I.J., 1975. Permeability to water in different adaptive phases of the same instar in the American house dust mite. Acarologia 17, 734–744.

Ellingsen, I.J., 1978. Oxygen consumption in active and quiescent protonymphs of the American house dust mite. J. Insect Physiol. 24, 13–16.

Glass, E.V., Yoder, J.A., Needham, G.R., 1998. Clustering reduces water loss by adult American house dust mites *Dermatophagoides farinae*. Exp. Appl. Acarol. 22, 31–37.

Hart, B.J., Le Merdy, L., 1987. Human dander-free house dust mite extracts. In: Mite Allergy, a World Wide Problem, Bad Kreuznach, September 1-2, 1987, pp. 47–49.

Knülle, W., 1961. Die Luftfeuchte-Unterschiedesempfindlichkeit der Mehlmilbe (Acarus siro). Z. Vgl. Physiol. 44, 463–477.

Knülle, W., 1962. Die Abhangigkeit der Luftfeuchte-Reaktionen de Mehlmilbe (Acarus siro) vom Wassergehalt des Korpers. Z. Vgl. Physiol. 45, 233–246.

Knülle, W., 1965. Die Sorption und Transpirtion des Wasserdampfes bei der Mehlmilbe (Acarus siro). Z. Vgl. Physiol. 49, 586–604.

Kuehr, J., Frischer, T., Karmaus, W., Meinert, R., Barth, R., Schraub, S., Daschner, A., Urbanek, R., Forster, J., 1994. Natural variation in mite antigen density in house dust and relationship to residential factors. Clin. Exp. Allergy 24, 229–237.

Matsumoto, K., Okamoto, M., Wada, Y., 1986. Effect of relative humidity on life cycle of the house dust mites, *Dermatophagoides farinae* and *D. pteronyssinus*. Jpn. J. San. Zool. 37, 79–90.

Molva, V., Nesvorna, M., Hubert, J., 2019. Feeding interactions between microorganisms and the house dust mite *Dermatophagoides pteronyssinus* and *Dermatophagoides farinae* (Astigmata: Pyroglyphidae). J. Med. Entomol. 56, 1669–1677.

Reka, S.A., Suto, C., Yamaguchi, M., 1992. Evidence of aggregation pheromone in the feces of house dust mite, *Dermatophagoides farinae*. Jpn. J. Sanit. Zool. 43, 339–341.

Sakaki, J., Suto, C., 1991. The termination of prolonged quiscence (diapause) in the house dust mite, *Dermatophagoides farinae* at room temperature. Jpn. J. Sanit. Zool. 42, 93–97.

Sakaki, J., Suto, C., Ito, H., 1990. Studies on the occurrence and termination of quiescence in nymphs of *Dermatophagoides farinae*. Jpn. J. Sanit. Zool. 41, 227–234.

Solarz, K., Skubala, P., Wauthy, G., Szilman, P., 2016. Body size variability in different forms of heteromorphic males in populations of the house dust mite *Dermatophagoides farinae* Hughes 1961 (Acari: Astigmata: Pyroglyphidae). Ann. Zool. 66, 329–336.

Suto, C., Sakaki, J., 1990. Studies on the factors influencing the induction, persistance and termination of prolonged quiscence (diapause) in house dust mite, *Dermatophagoides farinae*. Jpn. J. Sanit. Zool. 41, 375–381.

Suto, C., Sakaki, I., Ito, H., 1991. Comparative studies on the population dynamics of house dust mites, *Dermatophagoides farinae* and *D. pteronyssinus* in homes in Nagoya. Jpn. J. Sanit. Zool. 42, 129–140.

Walzl, M.G., 1992. Ultrastructure of the reproductive system of the house dust mites *Dermatophagoides farinae* and *D. pteronyssinus* (Acari, Pyroglyphidae) with special remarks on spermatogenesis and oogenesis. Exp. Appl. Acarol. 16, 85–116.

Wharton, G.W., Furumizo, R.T., 1977. Supracoxal gland secretions as a source of fresh water for Acridei. Acarologia 19, 112–116.

CHAPTER FIVE

Surveys of the occurrence of dust mites and their allergens

5.1 Associations with home characteristics

Let us flip the calendar back a few decades. Imagine you are living in the early 1970s. Not so long ago the involvement of house dust mites in the production of house dust allergens has been discovered. Imagine you are working for a public funding agency that, among other things, financially supports scientific research in the medical and biological fields. On your desk is an application for funding a research project. The applicants want to do an investigation into the factors that influence the occurrence of house dust mite allergens. They propose to collect samples of house dust from a large number of houses and determine the concentration of mite allergens in it. It is already known that these concentrations can be very different in different houses. Apart from taking dust samples, they will record the home characteristics that possibly influence the accumulation of mite allergens. A long list of characteristics are specified in the proposal: the number of inhabitants, age of the house, type and age of the furniture and carpeting, presence of pets, the regime of cleaning, airing, and heating, etc. Measurements will be made of temperature and humidity. The data will be analyzed using sophisticated statistical techniques (there is a mathematician in the team). The objective is of course to find out which properties of dwellings either enhance or suppress the occurrence of mite allergens. You don't have to make the decision about the funding of this proposal all by yourself. A board of advisors are at your disposal—experienced professionals in relevant fields of science from all over the world. They will give you their comments anonymously and with elaborate explanations.

But what would you think of it yourself? Does the proposal appeal to you? The reason why I am asking this is the following: It seems to have been exceptionally easy to convince funding organizations of the desirability of this kind of research. I infer this from the amazing number of studies of this

House Dust Mites
https://doi.org/10.1016/B978-0-443-19111-4.00014-9

kind that have been done and the results of which have appeared in the literature. My efforts to organize in an orderly manner all the information that was obtained from these investigations resulted in Table 5.1B, with a summary of the findings in Table 5.1A. The first part of this table lists characteristics that are associated with high mite numbers or allergen concentrations; the second part with low numbers and concentrations.

Table 5.1A Findings from studies concerning the associations between home characteristics and mite numbers.

The following home characteristics were found to be associated with <u>elevated</u> allergen levels or with <u>high</u> mite numbers

A	1	High indoor air humidity (3×)
A	2	Lower room temperature (4×)
A	3	Signs of dampness such as wet, moldy spots on walls or window condensation (4×)
A	4	Central heating (1×)
A	5	Use of forced air heating (1×)
A	6	Use of a tumble dryer or other drying method as opposed to using an outside line to dry the washings (1×)
A	7	Use of gas oven or hob (1×)
A	8	No extractor fan in the kitchen (1×)
A	9	Absence of mechanical ventilation (1×)
A	10	Low story level (3×)
A	11	Concrete bedroom floor (1×)
A	12	Absence of outer cavity wall (1×)
A	13	Absence of floor insulation (1×)
A	14	Weather board walls (1×)
B	15	Carpeted floors as opposed to smooth floors (2×)
B	16	Older homes as opposed to newer homes (4×)
B	17	Older carpet (3×)
B	18	Older mattresses (4×)
B	19	More heavily used fabric/upholstered furniture and carpeted floor areas (1×)
B	20	Old vacuum cleaner (1×)
B	21	Higher weight of sampled dust (1×)
C	22	Presence of pets (2×)
C	23	Higher number of residents (5×)
C	24	Single family homes as opposed to multiple family homes (1×)
C	25	Wooden detached houses as opposed to concrete apartment houses (1×)
C	26	Use of a cover or underblanket (2×)
C	27	Use of a blanket of animal hair (1×)
C	28	Use of feather pillows (1×)
C	29	Use of blankets (1×)
C	30	Bed with kapok or innerspring mattresses as opposed to foam mattresses (1×)

Table 5.1A Findings from studies concerning the associations between home characteristics and mite numbers—cont'd

The following home characteristics were found to be associated with elevated allergen levels or with high mite numbers

C 31 Use of wool underlays (1×)
C 32 Bedding washed at high temperature (1×)
C 33 High educational level of the inhabitants (1×)

The following home characteristics were found to be associated with low allergen levels (or mite numbers)

A 101 Central heating (1×)
A 102 Underfloor heating (1×)
A 103 Open fire in the living room (1×)
B 104 "Regular treatment" of the mattress (1×)
C 105 Higher number of residents (1×)
C 106 Smokers in the house (1×)
C 107 Interior sprung mattresses (1)

In brackets: the number of studies in which this association was reported.
First column: categories A, B, and C:
A: Association with or effect on temperature and humidity.
B: Association with or effect on food supply for dust mites.
C: Association with the behavior of the residents such as airing, heating, and cleaning.
Second column: the numbers used in Table 5.1B to indicate which findings were found in which studies.

Table 5.1B Studies concerning the association between home characteristics and high mite numbers or allergen concentrations.

Study	Location	Sampled sites measured concentration	N	Findings	Remarks
Charlet et al. (1977)	Bogota (Colombia)	Mattress and bedroom floor Mites/mL	20	16, 22	Mite population densities were higher in bedrooms with wooden floors as opposed to carpeted floors
Arlian et al. (1982)	Dayton and Cincinnati (Ohio, USA)	Mattress, bedroom floor, family room floor, and furniture Mites/g	19	15, 19	Mattresses were not found to be the major foci for mites

Continued

Table 5.1B Studies concerning the association between home characteristics and high mite numbers or allergen concentrations—cont'd

Study	Location	Sampled sites measured concentration	N	Findings	Remarks
Kuehr et al. (1994)	Freiburg, Loerrach/ Weil, Kehl (Germany)	Mattress _______ Dp1 and Df1	1291	For Der p1: 1, 2, 3, 10, 26, 27, 102, 105 For Der f1: 23, 18, 21, 26, 33, 104, 107	
van Strien et al. (1994)	Netherlands	Living room floor, bedroom floor, mattress _______ Der p1	516	1, 2, 3, 9, 12, 13, 15, 16, 17, 18, 23	"Mattresses in homes with continuous mechanical ventilation had almost twice lower levels (of Der p1) than mattresses in homes with natural ventilation"
Chan-Yeung et al. (1995)	Vancouver and Winnipeg (Canada)	Mattress and floor _______ Der p1 and Der f1	120	5, 16, 23, 24, 28	In both cities both allergens are found, but always more Der p1 than Der f1. In Winnipeg more Derp f1 is found than in Vancouver
Konishi and Uehara (1995)	Japan	Comforters, futons, pillows, carpets, tatamis _______ Antigens of D. p. and D.f.	72	25	
Barnes et al. (1997)	Barbados	Mattress, bedroom floor, family room floor, and furniture	17	23	Season (wet or dry) had no clear influence on the allergen levels

Table 5.1B Studies concerning the association between home characteristics and high mite numbers or allergen concentrations—cont'd

Study	Location	Sampled sites measured concentration	N	Findings	Remarks
		Der p1 and Der f1			
Wickens et al. (1997b)	Wellington (New Zealand)	Living room floor, bedroom floor, bedding	474	1, 15, 17, 23, 30, 31	A smooth floor as opposed to carpet seems to influence Der p1 in bed
		Der p1			
Couper et al. (1998)	Tasmania	Bedroom floor	72	2, 3, 6, 23	
		Der p1 and Der f1			
Luczynska et al., 1998	Norwich (UK)	Living room floor, bedroom floor, mattress	158	3, 7, 8, 10, 11, 16, 17, 18, 20, 29, 32, 103, 106	
		Der p 1			
Dharmage et al., 1999	Melbourne (Australia)	Bed and bedroom floor	485	1, 3, 4, 14, 15, 16, 18	Central heating is associated with higher Der p1 levels, also after correction for certain alleged confounders.
		Der p1			
Gross et al., 2000	Erfurt and Hamburg (Germany)	Floor and mattress	405	1, 2, 10, 17, 18, 22, 101	Indoor determinants of Der p1 and Der f1 concentrations are different. Most associations concern only Der p1. Presence of a dog was only associated with Der f1.
		Der p1 and Der f1			

N: sample size, i.e., number of houses or mattresses.

D.p. = *Dermatophagoides pteronyssinus*.

D.f. = *Dermatophagoides farinae*.

Findings (column 5): numbers refer to the home characteristics specified in Table 5.1A.

Table 5.1B lists the studies, ordered according to the year of publication, i.e. the earliest investigations first. In the earlier studies the number of mites per gram of house dust was taken as a measure of allergen exposure in a home. Soon, however, the mite counts were replaced by immunochemical assays that detect and quantify mite allergens in house dust. Most investigators switched to these newer techniques once they were available. Nobody questioned the superiority of these methods. After all, allergens are the direct cause of allergy symptoms. However, an understanding of the population dynamics of house dust mites is the starting point for an understanding of the accumulation of mite allergens in houses. The concentration of mite allergens is a derived indicator of the mite population density. It gives no information at all about the population structure and population development. Thus the switch from mite counts to allergen quantification was an unfortunate one for the acarologists.

Even brand new materials sometimes contain rather high concentrations of dust mite allergens (van der Hoeven et al., 1992; de Boer, 2002). It is very unlikely that these allergens were produced in situ by a mite population living in that item. Also the finding of, for instance, cat allergens in materials that were never in direct contact with a cat should make us careful and suspicious about any interpretation of reported levels of mite allergen. It is good to keep this in mind while reading the following account.

I tried to categorize the findings of the various studies, according to how they might be explained (A, B, or C in column 1 of Table 5.1A). To category A, I assigned all findings that might be explained on the basis of favorable or unfavorable humidity or temperature conditions. The findings 1 and 2 concern direct observations of temperature and humidity. Not many of these studies report such observations, but a lot of other studies have focused primarily on the association with indoor air humidity, ignoring all other home characteristics. These studies I listed in Table 5.2.

For statistical evaluations, observations must be quantified. But even then it is difficult. It is impossible to quantify the pattern of changes in RH and temperature by a single parameter. Some investigators took only one measurement of RH and temperature on the day the home was visited. Others registered the average over, for instance, 1 week. This may be a week during the spring, when mites increase their numbers, or in the autumn, when mite numbers are booming, or in the winter, when mites are hardly surviving.

Table 5.2 Studies concerning the association between air humidity and high mite numbers or allergen concentrations.

Study	Location	Sampled sites measured concentration	N	Remarks
Korsgaard (1982)	Aarhus (Denmark)	Mattress, living room, bedroom ——— Mites/g	98	Only intact mite bodies were counted. High absolute humidity of indoor air was related to greater concentrations of house dust mites
Korsgaard (1983a, 1983b)	Aarhus (Denmark)	Mattress, living room, bedroom ——— Mites/g	50	A high correlation between indoor air humidity and the occurrence of house dust mites exists in the colder season only
Harving et al. (1993)	Aarhus (Denmark)	Mattress, bedroom ——— Mites/g	96	A positive correlation was found between indoor air humidity ($P < .01$) and an inverse correlation ($P < .027$) between HDM concentration and indoor air exchange
Brown et al. (1995)	Melbourne (Australia)	Carpet and furniture ——— Living mites (on adhesive tape)	3	Mite numbers decreased over winter to very low levels in the living room but not in the bedroom, probably because bedrooms are rarely heated
Bigliocchi and Maroli (1995)	Rome (Italy)	Mattress, living room, bedroom ——— Living mites	90	Positive correlation of RH with mite numbers: $r = 0.89$, $P = .02$; 12.2% of houses without any mites. Mattress was most often infested
Sundell et al. (1995)	Stockholm (Sweden)	Mattress, bedroom ——— Der p1, Der f1 and Der m1	30	Association of high allergen concentrations with greater difference in absolute humidity between indoor and outdoor air and with low ventilation rate (measured with tracer gas)
Martin et al. (1997)	Christchurch (New Zealand)	Living room floor, bedroom floor, bed ——— Der p1	93	Association of elevated Der p1 levels with high RH

N: sample size, i.e., number of houses or mattresses.

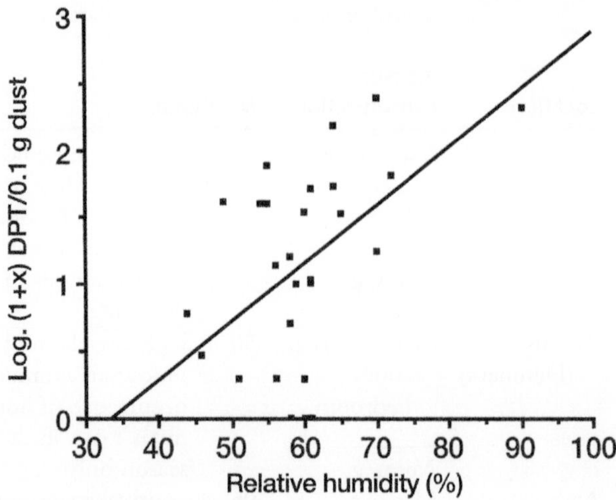

Fig. 5.1 Number of mattress *D. pteronyssinus* mites found at different bedroom humidities. Single humidity readings (using a whirling hygrometer) were taken on the day of sampling. *(From Hart, B.J., Whitehead, L., 1990. Ecology of house dust mites in Oxfordshire. Clin. Exp. Allergy 20, 203–209.)*

In any case only a very rough indication of the moistness of the indoor climate was obtained and used for the statistical evaluation, whereas, undoubtedly, the whole pattern of changes of physical conditions during the day and during the year is important for dust mites. In spite of all this, a positive correlation between the moistness of a dwelling and the concentration of mite allergen was observed whenever it was looked for. Fig. 5.1. is only one example. The association between RH and mite numbers is only marginally significant, but there are plenty more studies that show a positive correlation between RH and mite or allergen concentrations. That knocks down the cynics, including me. Also home characteristics that promote the moisture content of indoor air, such as poor ventilation (finding 8 and 9 in Table 5.1) or habitually drying the laundry inside the house (finding 6), were found to be associated with increased mite allergen levels.

Theoretically a rise of the temperature may work out positive as well as negative. If the absolute humidity is low the RH may drop below the critical (CEH) level, which is always unfavorable and lethal if it lasts long enough. On the other hand, if the absolute humidity is high, a rise of temperature may enhance mite growth and reproduction, provided there is enough food for them. In reality temperature and humidity conditions vary over time, and a single measurement or the average of several measurements is insufficient for a good understanding. Lower temperatures measured in the house, as well as home characteristics that lower the temperature in the mite habitats, such as the absence of floor insulation (finding 13) or absence of underfloor heating (finding 102), were found to be associated with elevated mite allergen levels. The temperature on the ground floor is generally lower than on higher story levels, certainly in temperate climates during the winter. This may explain the finding (nr 10) of higher allergen concentrations on lower story levels. Central heating is in some studies found to be associated with elevated allergen levels (finding 4), but in other studies the opposite was found (finding 101).

To category B (Table 5.1B), I assigned all findings that might be explained on the basis of food supply for dust mites. These include also the age of the sampled carpet or mattress, because food accumulates over time.

To category C, I assigned all findings that might be explained by the behavior of the residents such as airing, heating, and cleaning. Also some other findings that I found totally inexplicable with the present state of our knowledge, I included in this category.

Often a correction was made for alleged confounding factors. But, no doubt, some confounders are not recognized. For instance, a significant positive association ($P < .001$) was found to exist between the educational level of the inhabitants of a house and the concentration of allergen (finding 33). Clearly, the influence of the educational level of the inhabitants on the accumulation of mite allergen must be an indirect one. So we must be missing one or a few confounders. The finding that smokers in the house have a negative influence on allergen levels (finding 106) is another challenge for our imagination.

Couper et al. (1998) also did a rather unexpected observation. These are their own words: "unmeasured confounders such as thickness and type of carpet and the use of carpet in other rooms, may partially explain the unexpected finding that carpeted bedrooms which were vacuumed at least once a week had higher allergen levels compared with those vacuumed less often. An alternative explanation is that vacuuming decreased the numbers of (predatory) *Cheyletus* (mites)."

Kuehr et al. (1994) determined concentrations of allergens from both *D. pteronyssinus* (Der p1) and *D. farinae* (Der f1). The associations of home characteristics with elevated allergen concentrations turned out to be quite different for the two species, as can be seen in Table 5.1. Of the 15 associations that could be classified as statistically significant ($P < .05$), only one (nr. 26) was true for both Der p1 and Der f1. Particularly noteworthy is the finding that the number of residents per square meter shows a significant *positive* association with the concentration of Der f1, but a significant *negative* association with the concentration of Der p1. These findings confirm that *D. pteronyssinus* and *D. farinae* have different ecological requirements. But we have a long way to go before we can explain each one of these associations.

Not every association is difficult to explain. Mites need moisture, food, and shelter. Carpets accumulate food for mites and provide them with shelter. Mites do not live on a smooth, clean surface. It is quite easy to remove all mites, all their food, and all their allergens from a smooth floor, in sharp contrast with a carpet. At first sight it may be surprising to find high allergen concentrations per gram of dust or per square meter from a smooth floor. But don't forget that the dust taken from a carpet is only a small fraction, whereas the dust taken from a smooth floor may be practically the total amount that was present. And, secondly, don't forget that the dust taken from a smooth floor must have been brought there from elsewhere—either elsewhere in the same house or from some other house. The likeliest vehicles are the clothes of people.

These lavishly funded research projects did not bring forth a deeper understanding of dust mite ecology. Some houses had very low allergen levels. Nobody tried to pin down the exact reason for this by an in-depth follow-up study of these habitats. Was there a viable dust mite population? Was the microclimate suitable for dust mites? Opportunities were ignored.

5.2 Density-dependent and density-independent factors

Air humidity is very important for dust mites. But, even so, the correlation between home dampness and mite allergen concentrations that we see again and again is not easy to explain. If a change in humidity affects the rate of population growth, it does so in the same way, no matter if the density is 500 or 10,000 mites per square meter. As long as no factors like food scarcity or predation pressure come into play, the mite population increases exponentially, i.e., faster and faster. Even under conditions of marginal humidity, which would allow only very slow reproduction, the population increase will eventually become very rapid because the number of reproducing females continues to increase. A limit to the population density is set by such factors as the availability of food or, possibly, the development of a predator population (*Cheyletus*) but not by humidity. Nevertheless, a positive correlation between air humidity and the numbers of mites or the concentration of mite allergen was found whenever it was looked for.

Several possible, not mutually exclusive explanations present themselves:

(1) The allergen concentration in house dust is the result of a balance between the amount that is removed and the amount that is synthesized by the local house dust mite population every day. The rate of allergen production is higher at higher humidities. However, the removal rate is independent of humidity. Consequently, the equilibrium settles at a higher level when RH is high.

(2) If most mite populations encountered inside the home are still developing, the upper limit of the population density is usually not reached. In moist homes, however, the population development is generally more advanced.

(3) Air humidity may also indirectly influence mite populations, through its effect on microbes such as fungi. Microbes may be important for processing the skin scales that serve as the main food for dust mites or synthesize valuable nutrients for dust mites. Conceivably, when humidity is too low for fungus growth, suitable food for dust mites is not available or is in short supply. This possibility is discussed in Section 9.3. and 9.4.

5.3 Are there any houses without mites?

It is possible that climatic conditions are so extreme that even inside houses no mites can live. Lang and Mulla (1977) sampled 15 houses in each of four climatic zones in southern California. Live mites were not found in the 15 houses sampled in the desert, in contrast to the houses on the coast and in an inland valley. In a survey of Saudi Arabia (Al-Frayh et al., 1993) no house dust mite antigen was found in homes in the desert, whereas fairly high levels were measured in coastal and mountainous regions. Also the influence of altitude is quite clear. A negative correlation between mite numbers in house dust and elevation was found by Spieksma (1967) in the Swiss Alps, by Ordman (1971) in South Africa and by Auer et al. (1985) in the Austrian Alps. Auer et al. took dust samples from houses in the mountainous Austrian state of Tirol, at elevations ranging from 536 to 2150 m above sea level (Table 5.3). Higher than 1300 m above sea level the mite numbers never exceeded 11 mites/100 mg of dust, whereas in the valleys below 600 m the usual high mite densities were recorded. Thus it seems that the atmospheric conditions at high altitudes are unsuitable for

Table 5.3 Mite densities in dust from houses at different altitudes in Tirol, Austria. (Auer et al., 1985).

Township	Altitude (m)	Pyroglyphidae (mites/100 mg)
Hochgurgl	2150	0, 1, 2
Hochsölden	2090	2
Kühtai	2020	2, 5, 11
Obergurgl	1927	3, 3
Vent	1900	5, 5
St. Christoph	1802	1, 2
Praxmar	1710	0, 0, 0
Mandarfsen	1670	0, 6
Wirl	1654	1, 1
Plangeross	1652	2
Haggen	1650	0, 2
Gies im Sulztal	1600	1, 3, 5
Galtür	1584	0, 1, 4
Niederthai	1550	0, 1, 2
Lechleiten	1539	0, 6
Kaisers	1518	0, 1
St. Siegmund	1516	0, 0, 0
St. Veit	1495	1, 2, 3
Obertilliach	1450	0, 4, 7
Serfaus	1427	0, 1, 2
Innervillgraten	1408	0, 2, 2
St. Jakob	1398	0, 2, 3
Sölden	1377	0, 0
St. Anton	1300	1, 7
St. Leonhard	1200	0, 1, 81
St. Jodok	1129	1
Steeg	1122	2, 4
Jerzens	1100	5, 10, 10
Mieders	982	58
Absam	632	0, 75
Ampass	630	410
Zirl	620	6
Innsbruck	574	4, 9, 14, 31, 58
Hall	562	12, 41
Schwaz	536	56, 145

house dust mites, also inside houses, also in beds. Relevant information about the evolutions of temperature and humidity in the mite's habitats was not registered. An interesting question is: What is the origin of the few mites that were found at high elevations? Even at 2150 m above sea level one or a few mites were found. Do these mites represent a dwindled but self-sustaining population? Or do they represent the mites that are transported on the clothes of people who visited mite-infested houses down in the valley? In connection with these questions, the study by Moyer et al. (1985) in Denver (Colorado, USA) is very interesting. The city of Denver is built between 1560 and 1730 m above sea level. The relative humidity averages between 40% and 50% with little seasonal variation. During the heating season, indoor relative humidity tends to be even lower except in homes with humidifiers where it is about 50%. In the Abstract of the paper by Moyer et al. We read: "Sixty-four samples of house dust from 16 long-established households in the Denver, Colorado area were analyzed for the presence of house dust mites. No mites were found in house dust from 12 of the sampled houses and small numbers (10–40 mites/g of house dust) were found in the other four. In an additional four houses which contained furniture recently imported from other areas, 100 to 360 mites/g of dust were found, and 2 years later up to 200 mites/g were still present. Twenty-eight percent of the mites in repeat collections from the latter homes were alive." Further in the article we read: "Repeat sampling was done of six areas in the four houses which 'imported' mites at 6-month intervals. These houses showed a persistence of mites during the two years of collection. To make sure that this did not merely reflect a 'mite graveyard,' collected samples were stored in a cool, humid environment and analyzed within 48 h. Approximately one-third of the mites in these samples were still alive, indicating true persistence. Most of the mites in the native Denver samples were immature, and no species identification could be made. All 14 of the mites initially found in the imported furniture on which species identification were made were *D. pteronyssinus*, reflecting the previous residence in areas of high relative humidity. In later collections from these houses, both *D. pteronyssinus* and *D. farinae* were present in approximately equal numbers. Perhaps this indicates selective persistence of what had originally been a small minority of the latter species, which is known to be better adapted to dry conditions."

It is amazing that a mite population could survive for such a long time in a climate that is clearly not conducive for dust mite population growth. To explain this finding the possibility should be considered that atmospheric conditions influence mite growth also indirectly, namely through its influence on the development of a suitable food supply. The imported furniture contained not only live mites but also a food source for them. Perhaps a source of suitable food for dust mites for some reason cannot develop in the high altitude atmospheric conditions. And food may be more important than humidity.

In less extreme climatic zones, houses without mites are rare. Arlian et al. (1992) sampled 252 houses all over the United States. All homes in all geographic areas were positive for house dust mites. This finding surprised me. Not a single mite-free house in such a big sample. But all these homes were the homes of allergy patients, which may explain the absence of mite-negative houses.

In my not so very long career as a house dust mite investigator, I have a number of times encountered an occupied dwelling where a large sample of dust from an old carpet did not contain a single mite. I can distinctly remember at least three cases. I was surprised, unable to explain the observation, but I did not give it much thought at the time. I did not analyze the dust for allergen content, but detectable quantities of allergen may have been present as the result of contamination with allergen from sources elsewhere. Contamination with allergens commonly occurs as is revealed by the presence of detectable quantities of allergen in brand new carpets and brand new mattresses. The clothes of people working in the factory where the carpets and mattresses were made must have been the vehicle. Therefore the existence of houses without a viable mite population may often remain unnoticed in studies that rely on the determination of allergen concentrations. I speculate that the houses where very low mite allergen concentrations are found belong to a group of houses that, for some reason, are unsuitable for house dust mites. It is a pity that the presence of live mites in these houses was not checked. Even more interesting would it be to try to rear mites with dust from these houses as a rearing medium or on fragments of old carpets from these houses. Is there an adequate food supply for mites there? The possibility that abiotic conditions in a house effect the dust mite population also indirectly, through the food chain, has been studied inadequately so far.

5.4 Utility buildings and public places

As I argued in the previous section, the absence of mites is often more intriguing than their presence. Also utility buildings like offices, schools, hotels, and hospitals were often found to contain no mites and very low allergen concentrations in carpets and other dust collecting items. Table 5.4. lists the findings from investigations of utility buildings and public places. Particularly interesting are the studies where mite numbers were determined instead of allergen concentrations, especially those where private houses in the same area were also sampled (Sarsfield, 1974; Rao et al., 1975; Green et al., 1992; Brown, 1994). The amounts of allergen were generally very low in comparison with the concentrations that were found in private homes.

Table 5.4 Studies concerning the mite infestations in public places and utility buildings.

Author(s) study	location	Measure of allergen conc. in dust	Findings
Sesay and Dobson (1972)	Scotland (UK)	Mites	Only few mites in hospital beds
Sarsfield (1974)	Leeds (UK)	Mites	No mites were found in any of 15 hospital ward mattresses. In contrast, all 10 domestic mattresses were infested
Rao et al. (1975)	Cardiff (UK)	Mites	100 hospital beds were free of mites, in contrast to 50 private mattresses
Dybendal et al. (1989)	Norway		12 schools; carpets relatively free of mites
Friedman et al. (1992)	Connecticut River Valley (USA)	D.p and D.f. antigens	15 homes and 23 workplaces very low levels of allergen in workplaces
Green et al. (1992)	Sydney (Australia)	Mites and allergen	38 public places had on average 59 mites/g. Bedding and furniture in 18 private houses had 229 mites/g and the floors had 274 mites/g on average. The authors suggest that air-conditioning in the public buildings may be an important factor; however, several of the schools were not air-conditioned and yet their allergen levels were mainly low
Brown (1994)	Melbourne (Australia)	Mites (on tape)	Carpet in an office building was free of mites in contrast to 3 homes
Babe et al. (1995)	New Castle County, Delaware (USA)	Mites	No mites or only insignificant numbers in 20 carpeted hospital hallways and 20 carpeted patient rooms
Einarsson et al. (1995)	Linkoping (Sweden)	Der p1 and Der f1	4 schools, 29 classrooms Mite allergen concentrations were significantly ($P < .0001$) higher in dust from tables than in dust form chairs

Continued

Table 5.4 Studies concerning the mite infestations in public places and utility buildings—cont'd

Author(s) study	location	Measure of allergen conc. in dust	Findings
			and the floor. Authors think that it may have been transported in the hair of the children. Mite allergen concentrations were significantly ($P < .05$) higher in the only school with a moisture problem
Janko et al. (1995)		Der p1	14 offices; four had $>1\,\mu g/g$ "In all cases the infestation of dust mites was localized to a few specific work areas"
Munir et al. (1995)	Sweden	Mite allergen	In 7 day care centers a total of 22 sections were sampled. Mite allergen was found in 9 sections but not in any dust sample from the (smooth) floors
Zock and Brunekreef (1995)	Rotterdam and small towns in eastern Holland	Der p1	18 schools with smooth floors and 31 with carpeted floors were sampled. Carpeted floors had higher Der p1 levels than smooth floors, but these levels were considerably lower than in dust from homes. Age of the floor cover, the number of classrooms, and the presence of damp spots were related to Der p1 concentrations
Custovic et al. (1998)	NW–England, London and N-Wales (UK)	Der p1	14 hospitals (9, 4, and 1, respectively) carpets, mattresses, and upholstered chairs; levels of Der p1 were low
Custovic et al. (1994)	Manchester (UK)	Der p1	5 schools, 6 hotels, 4 cinemas, 6 pubs, 3 buses, 2 trains, and 12 domestic households: Der p1 concentration was

Table 5.4 Studies concerning the mite infestations in public places and utility buildings—cont'd

Author(s) study	location	Measure of allergen conc. in dust	Findings
			significantly higher in the private homes than in comparable sites in public places except for cinema seats
Wickens et al. (1997a)	Christchurch and Wellington (New Zealand)	Der p1	Dust was obtained from hotels, hospitals, rest homes, churches, primary schools, childcare centers, offices, and airplanes. Der p1 levels were much lower in these public places than in domestic dwellings
Konishi and Uehara (1999)	Kobe (Japan)	Mite antigen	Many dust samples were taken from each of 4 hospitals, 2 hotels, 2 ryokans, 1 cinema, and 4 offices. Elevated allergen levels were recorded in many cinema chairs and floors of ryokans. Only 2% of hospital mattresses and 5% of hospital floors had elevated allergen levels. The air-conditioned offices had very low levels

References

Al-Frayh, A.R., Hasnain, S.M., Gad-El-Rab, M.O., Al-Mobairek, K., 1993. Unique variation in house dust mite contents in Saudi Arabia. Allergy 48 (Suppl. 16), 108.

Arlian, L.G., Bernstein, I.L., Gallagher, J.S., 1982. The prevalence of house dust mites, *Dermatophagoides* spp., and associated environmental conditions in homes in Ohio. J. Allergy Clin. Immunol. 69, 527–532.

Arlian, L.G., Bernstein, D., Bernstein, I.L., Friedman, S., Grant, A., Lieberman, P., Lopez, M., Mezger, J., Platts-Mills, T., Schatz, M., Spector, S., Wasserman, S.I., Zeiger, R.S., 1992. Prevalence of dust mites in the homes of people with asthma living in eight different geographic areas of the U.S. J. Allergy Clin. Immunol. 90, 292–300.

Auer, H., Demetz, H., Frank, A., Janetschek, H., 1985. Untersuchung über das Vorkommen der Hausstaubmilbe in Tal- und Hochlagen Tirols als Voraussetzung einer Allegenkarenz-Empehlung (in German). Allergologie 8, 123–127.

Babe, K.S., Arlian, L.G., Confer, P.D., Kim, R., 1995. House dust mite (*Dermatophagoides farinae* and *Dermatophagoides pteronyssinus*) prevalence in the rooms and hallways of a tertiary care hospital. J. Allergy Clin. Immunol. 95, 801–805.

Barnes, K.C., Fernandez-Caldaz, E., Trudeau, W.L., Milne, D.E., Brenner, R.J., 1997. Spatial and temporal distribution of house dust mite (Astigmata: Pyroglyphidae) allergens Der p I and Der f i in Barbadian homes. J. Med. Entomol. 34, 212–218.

Bigliocchi, F., Maroli, M., 1995. Distribution and abundance of house dust mites (Acarina: Pyroglyphidae) in Rome, Italy. Aerobiologia 11, 35–40.

Brown, S.K., 1994. Optimisation of a screening procedure for house dust mite numbers in carpets and preliminary application to buildings. Exp. Appl. Acarol. 18, 423–434.

Brown, S., Cole, I., Martin, A., 1995. The effect of building factors and indoor climate on house dust mite numbers in three houses. In: Tovey, E., Finfoot, A., Sieber, L. (Eds.), Mites Asthma and Domestic Design II; University Printing Service. Univ. Sydney, pp. 92–95.

Chan-Yeung, M., Becker, A., Lam, J., Dimich-Ward, H., Ferguson, A., Warren, P., Simon, E., Broder, I., Manfreda, J., 1995. House dust mite allergen levels in two cities in Canada: effects of season, humidity, city and home characteristics. Clin. Exp. Allergy 25, 240–246.

Charlet, L.D., Mula, M.S., Sanchez-Medina, M., 1977. Domestic acari of Colombia: abundance of the European house dust mite, Dermatophagoides pteronyssinus (Acari: Pyroglyphidae) in homes in Bogota. J. Med. Entomol. 13, 709–712.

Couper, D., Ponsonby, A.L., Dwyer, T., 1998. Determinants of dust mite allergen concentrations in infant bedrooms in Tasmania. Clin. Exp. Allergy 28, 715–723.

Custovic, A., Taggart, S., Woodcock, A., 1994. House dust mite and cat allergen in different indoor environments. Clin. Exp. Allergy 24, 1164–1168.

Custovic, A., Fletcher, A., Pickering, C.A.C., Francis, H.C., Green, R., Smith, A., Chapman, M., Woodcock, A., 1998. Domestic allergens in public places III: house dust mite, cat, dog and cockroach allergens in British hospitals. Clin. Exp. Allergy 28, 53–59.

de Boer, R., 2002. Allergens, Der p 1, Der f 1, Fel d 1 and can f 1, in newly bought mattresses for infants. Clin. Exp. Allergy 32, 1602–1605.

Dharmage, S., Bailey, M., Raven, J., Cheng, A., Rolland, J., Thien, F., Forbes, A., Abramson, M., Walters, E.H., 1999. Residential characteristics influence Der p 1 levels in homes in Melbourne, Australia. J. Allergy Clin. Immunol. 29, 461–469.

Dybendal, T., Hetland, T., Vik, H., Apold, J., Elsayed, S., 1989. Dust from carpeted and smooth floors. I. Comparative measurements of antigenic and allergenic proteins in dust vacuumed from carpeted and non-carpeted classrooms in Norwegian schools. Clin. Exp. Allergy 19, 217–224.

Einarsson, R., Munir, A.K.M., Dreborg, S.K.G., 1995. Allergens in school dust: II. Major mite (Der p I, Der f I) allergens in dust from Swedish schools. J. Allergy Clin. Immunol. 95, 1049–1053.

Friedman, F.M., Friedman, H.M., O'Connor, G.T., 1992. Prevalence of dust mite allergens in homes and workplaces of the upper connecticut river valley of New England. Allergy Proc. 13, 259–262.

Green, W.F., Marks, G.B., Tovey, E.R., Toelle, B.G., Woolcock, A.J., 1992. House dust mites and mite allergens in public places. J. Allergy Clin. Immunol. 89, 1196–1197.

Gross, I., Heinrich, J., Fahlbusch, B., Jäger, B., Bischoff, W., Wichmann, H.E., 2000. Indoor determinants of Der p 1 and Der f 1 in house dust are different. Clin. Exp. Allergy 30, 376–382.

Harving, H., Korsgaard, J., Dahl, R., 1993. House dust mites and associated environmental conditions in Danish homes. Allergy 48, 106–109.

Janko, M., Gould, D.C., Vance, L., Stengel, C.C., Flack, J., 1995. Dust mite allergens in the office environment. Am. Ind. Hyg. Assoc. J. 56, 1133–1140.

Konishi, E., Uehara, K., 1995. Distribution of Dermatophagoides mite (Acari: Pyroglyphidae) antigens in homes of allergic patients in Japan. Exp. Appl. Acarol. 19, 275–286.

Konishi, E., Uehara, K., 1999. Contamination of public facilities with *Dermatophagoides* mites in Japan. Exp. Appl. Acarol. 23, 41–50.

Korsgaard, J., 1982. Preventive measures in house dust allergy. Am. Rev. Respir. Dis. 125, 80–84.

Korsgaard, J., 1983a. Mite asthma and residency. Am. Rev. Respir. Dis. 128, 231–235.

Korsgaard, J., 1983b. House dust mites and absolute indoor humidity. Allergy 38, 85–92.

Kuehr, J., Frischer, T., Karmaus, W., Meinert, R., Barth, R., Schraub, S., Daschner, A., Urbanek, R., Forster, J., 1994. Natural variation in mite antigen density in house dust and relationship to residential factors. Clin. Exp. Allergy 24, 229–237.

Lang, J.D., Mulla, M.S., 1977. Distribution and abundance of house dust mites, *Dermatophagoides* spp., in different climatic zones of Southern California. Environ. Entomol. 6, 213–216.

Luczynska, C., Sterne, J., Bond, J., Azima, H., Burney, P., 1998. Indoor factors associated with concentrations of house dust mite allergen, Der p 1, in a random sample of houses in Norwich, UK. Clin. Exp. Allergy 28, 1201–1209.

Martin, I.R., Henwood, J.L., Wilson, F., Koning, M.M., Pike, A.J., Smith, S., Town, G.I., 1997. House dust mite and cat allergen levels in domestic dwellings in Christchurch. N. Z. Med. J. 110, 229–231.

Moyer, D.B., Nelson, H.S., Arlian, L.G., 1985. House dust mites in Colorado. Ann. Allergy 55, 680–682.

Munir, A.K.M., Einarsson, R., Dreborg, S.K.G., 1995. Mite (Der p I, Der f I), cat (Fel d I) and dog (Can f I) allergens in dust from Swedish day-care centres. Clin. Exp. Allergy 25, 119–126.

Ordman, D., 1971. The incidence of 'climate asthma' en South Africa: its relation to the distribution of mites. S. Afr. Med. J. 71, 739–740.

Rao, V.R.M., Dean, B.V., Seaton, A., Williams, D.A., 1975. A comparison of mite populations in mattress dust from hospital and from private houses in Cardiff, Wales. Clin. Allergy 5, 209–215.

Sarsfield, J.K., 1974. Role of house dust mites in childhood asthma. Arch. Dis. Child. 49, 711–715.

Sesay, H.R., Dobson, R.M., 1972. Studies on the mite fauna of house dust in Scotland with special reference to that of bedding. Acarologia 14, 384–392.

Spieksma, F.T.M., 1967. The House-Dust Mite Dermatophagoides Pteronyssinus (Trouessart, 1897), Producer of the House-Dust Allergen. Academic thesis, Rijks Universiteit Leiden, Leiden.

Sundell, J., Wickman, M., Pershagen, G., Nordvall, S.L., 1995. Ventilation in homes infested by house dust mites. Allergy 50, 106–112.

van der Hoeven, W.A., de Boer, R., Bruin, J., 1992. The colonisation of new houses by house dust mites (Acari: Pyroglyphidae). Exp. Appl. Acarol. 16, 75–84.

van Strien, R.T., Verhoeff, A.P., Brunekreef, B., van Wijnen, J.H., 1994. Mite antigen in house dust: relationship with different housing characteristics in the Netherlands. Clin. Exp. Allergy 24, 843–853.

Wickens, K., Martin, I., Pearce, N., Fitzharris, P., Kent, R., Holbrook, N., Siebers, R., Smith, S., Threthowen, H., Lewis, S., Town, I., Crane, J., 1997a. House dust mite allergen levels in public places in New Zealand. J. Allergy Clin. Immunol. 99, 587–593.

Wickens, K., Siebers, R., Ellis, I., Lewis, S., Sawyer, G., Tohill, S., Stone, L., Kent, R., Kennedy, J., Slater, T., Crothall, A., Threthowen, H., Pearce, N., Fitzharris, P., Crane, J., 1997b. Determinants of house dust mite allergen in homes in Wellington, New Zealand. Clin. Exp. Allergy 27, 1077–1085.

Zock, J.P., Brunekreef, B., 1995. House dust mite allergen levels in dust from schools with smooth and carpeted classroom floors. Clin. Exp. Allergy 25, 549–553.

CHAPTER SIX

Where do mites survive the winter in a temperate climate?

Here in Holland, and in many other countries in the temperate climatic zones, streets and roofs of houses look wet in winter much more often than in summer. That is because the temperature is closer to the dew point. After a rain shower during the winter it takes much longer before pavements and roofs of houses are dry again. But this apparent wetness of the winter season is deceptive. In fact the cold winter air generally contains less water vapor than the warm summer air. And inside houses, where the temperature is artificially kept comfortably warm by the heating system, the RH is generally lower than in summer. Several studies have shown that mite allergen concentrations peak in summer and decline in winter. Thus it is evident that domestic mites experience unfavorable conditions during the heating season. But somehow they survive. Are there any sites in a house where dust mites retreat to survive the winter?

6.1 Mattresses

Many of the skin scales that are shed by a person must end up in the bed. And when the person is in this bed, water vapor emanates from the body. Therefore I once believed that conditions in any bed that is regularly slept in must be suitable for a dust mite population. Now I know that this contention is definitely wrong. Auer et al. (1985) investigated dust mite populations in a mountainous part of Austria. Most of the dust samples were taken from beds. In houses at low altitudes the beds were usually infested by dust mites but at high altitudes beds were virtually free of mites. Also Korsgaard (1983) reported on a number of homes in Denmark where humidity was generally low and house dust mites accordingly scarce in all the items sampled; the beds were no exception. Similar findings were reported by Turos (1979) and Wickman et al. (1993). Hospital beds and beds in hotels have also been found to contain very low numbers of mites or levels of allergen. An interesting observation in this connection was done by Arlian et al. (1982), who did an extensive survey of 19 houses in Ohio (USA). They write: "The fact that mite abundance on carpet floors under and alongside beds was significantly

House Dust Mites
https://doi.org/10.1016/B978-0-443-19111-4.00016-2

higher than that in mattresses would indicate that the mites were breeding on the floors and were not migrating there or dispersing there from the mattresses." Also interesting is the observation by Hughes and Maunsell (1973): "One of the few beds we have examined which was found to be free from mites, was occupied by two people: one using it at night, the other during the day; so maintaining a dry atmosphere in the bed all the time."

Table 6.1 After the resident mite population was killed by freezing, cultured mites were introduced into fragments cut out of an old mattress. The number of mites recovered after four weeks at different depths in six columns, reveal the suitability of the fragments to support survival and reproduction of introduced *D. pteronyssinus*.

Layer	Distance to nearest surface (mm)	Column						Total
		1	**2**	**3**	**4**	**5**	**6**	
	0							
Ticking		2	7	0	0	9	172	**190**
	8							
1		85	8	9	3	132	246	**483**
	30							
2		0	1	4	3	14	17	**39**
	52							
3		0	3	0	0	2	5	**10**
	52							
4		2	2	66	5	3	18	**96**
	30							
5		37	105	347	82	139	68	**778**
	8							
Ticking		88	103	59	13	56	244	**563**
	0							

From de Boer, R., Kuller, K., 1994. House dust mites (*Dermatophagoides pteronyssinus*) in mattresses: vertical distribution. Proc. Exp. Appl. Entomol. N.E.V. Amsterdam 5, 129–130.

However, in other circumstances, beds are indeed the only place where mites can survive. In a survey of some Dutch cottages during the winter (van Bronswijk, 1973), live mites were found only in beds.

Investigations of mite populations in mattresses were done mostly by taking dust samples from the surface using a vacuum cleaner. For my investigations into the possibility to control mites in mattresses using an electric blanket, I dissected several mattresses. I found living mites deep inside the polyurethane core. I also examined the suitability of the mattress material to support a mite population. Columns were cut out of a 12-year-old mattress at six places. These columns were sliced horizontally into layers and any mites living inside were killed by freezing. Then 80 cultured *D. pteronyssinus* of various ages were placed on each fragment. The fragments were stored at 25°C and 75%RH for 4 weeks. Then the number of live mites in each fragment was determined by the heat escape method. The results are presented in Table 6.1. The number of mites found in each fragment probably reflects the amount of food in it. Of all mites 35% were recovered from the ticking. The slices of the polyurethane core were of equal thickness. Yet most mites were present in the outer layers (Friedman's test, $P < .01$, 4 d.f.), indicating that a suitable habitat is found also below the ticking, but not much deeper than 20 mm.

In order to monitor temperature and RH inside a mattress, a 'sword probe' was fitted underneath the ticking. Measurements were recorded every 5 min before, during, and after a man (me), wearing pajamas, lay down on his back on top of it, while covered by a duvet. Fig. 6.1. shows the evolutions of RH, temperature, and CEH underneath the center of the men's back. Ambient conditions and conditions in the unoccupied bed were, on this summer day, favorable for dust mites, i.e., RH > CEH. Much to my surprise this situation was reversed when the bed was occupied. The CEH value goes up when the temperature rises. The rise of the RH is damped by the rising temperature. The result is a combination of RH and temperature whereby mites lose body water (RH < CEH) in spite of the sweating body.

In order to monitor temperature and RH deep inside the foam core, a horizontal slit was made 4.5 cm underneath the upper surface of the mattress. The recorded temperatures and RH, one day in late summer, are shown in Fig. 6.2. Also on this day ambient conditions were good for mites (RH > CEH). At this depth inside the mattress, changes of temperature and RH due to the presence of a human body are much more gradual than near the surface. The temperature is not so much increased at this distance away from the man's body. Consequently, the CEH is not so strongly influenced and the rise of RH is not so strongly damped. Conditions remained favorable for mites all the time. But if the atmospheric air would have been just a little dryer so that ambient RH < CEH, then it is quite likely that the presence of the man's body would indeed have created good conditions for mites (RH > CEH) while they were unfavorable (RH < CEH) before. Unfortunately, I have no records to testify whether this truly happens in reality.

Only one more record is available of the evolutions of temperature and RH inside an occupied mattress. This one was made directly underneath the ticking on a cold and dry winter day and is shown in Fig. 6.3. As it is

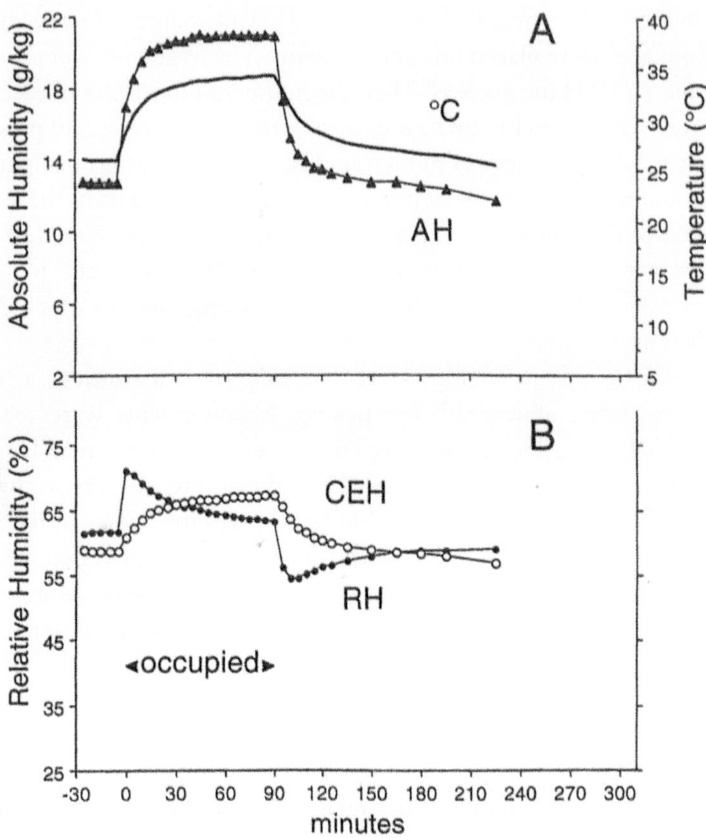

Fig. 6.1 Changes of temperature (°C), absolute humidity (AH), and relative humidity (RH) in an occupied bed. Recordings were taken directly underneath the ticking at the center of occupant's back on 14 July 1992. *(From de Boer, R., Kuller, K., 1997. Mattresses as a winter refuge for house dust mites. Allergy 52, 299–305.)*

quite often the case in winter in Holland, ambient conditions were unfavorable for mites (RH < CEH). Immediately after the man lies down on the mattress we see an upward spike in the RH curve. Water vapor penetrates into the mattress faster than warmth does. But soon the rising temperature causes the RH to go down again. The RH rose above the CEH for only a very short while, certainly insufficient time for the mites to restore water balance.

I never observed the RH in the occupied bed to rise above the CEH, whereas the ambient air was below it. However, it is quite likely that this situation would arise if the conditions shown in Fig. 6.3 were just a little different, with only a slightly higher ambient AH.

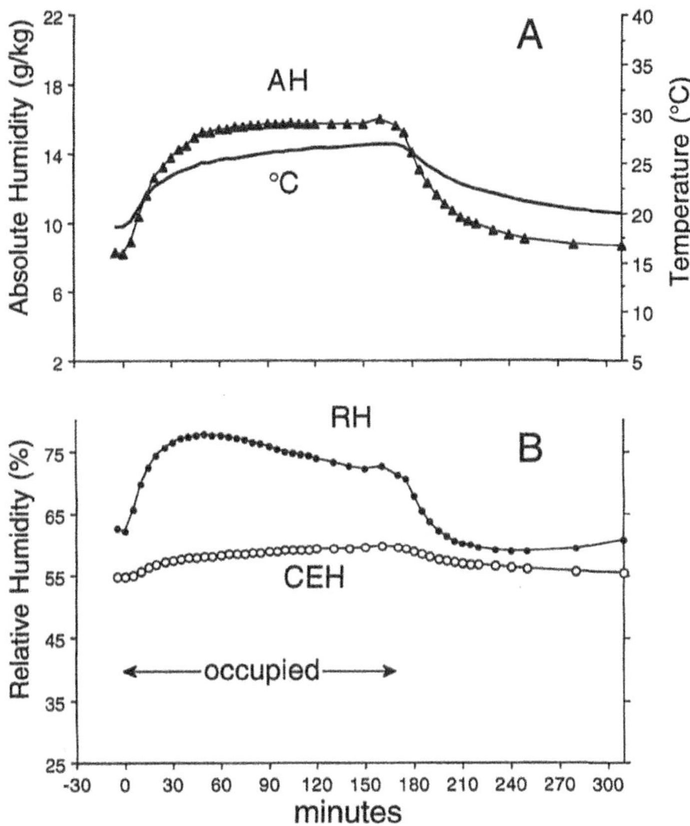

Fig. 6.2 Recordings were taken 4.5 cm below the surface of the mattress, at the center of occupant's back on 20 September 1992. *(From de Boer, R., Kuller, K., 1997. Mattresses as a winter refuge for house dust mites. Allergy 52, 299–305.)*

Thus the observation in several studies that beds can sometimes remain free of house dust mites in spite of the regular occupation may be explained by the concomitant rise of temperature when the bed is occupied. When the temperature rises, the mites require more moisture—more sometimes than the amount that is supplied by the occupant's body. On the other hand, it is conceivable that situations occur where house dust mites can survive the winter period only in beds or in stuffed chairs. But such situations are less common than I once believed.

The distribution of the mite fauna in upholstered furniture presumably resembles that in mattresses because mattresses and upholstery are quite similar regarding the aspects that are important for dust mites.

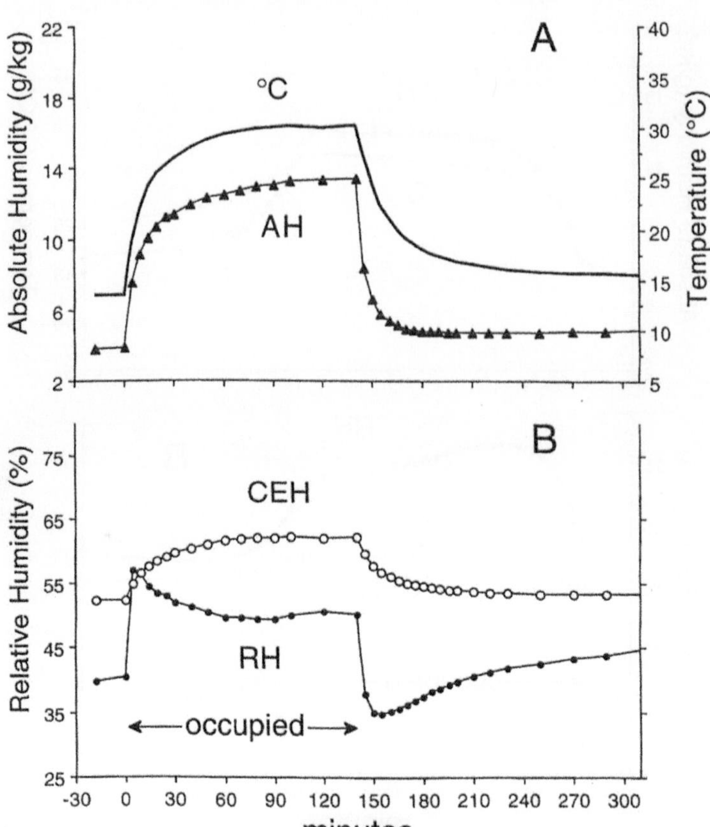

Fig. 6.3 Recordings were taken directly underneath the ticking at the center of occupant's back on 22 February 1992. *(From de Boer, R., Kuller, K., 1997. Mattresses as a winter refuge for house dust mites. Allergy 52, 299–305.)*

> ## 6.2 Carpeted floors and rugs

For one of my investigations I used data loggers to make continuous records of temperature and RH inside rugs on the ground floor of several houses. Two examples are shown in Figs. 6.4 and 6.5. The first one shows a record on four consecutive days in winter. The RH rises above the CEH each day but for only a few hours. The air we exhale contains more water vapor than the inhaled air. Thus breathing of the inhabitants, but also household activities like cooking, bathing, and doing the laundry bring extra water vapor into the air, especially during the day, whereas by day ventilation is

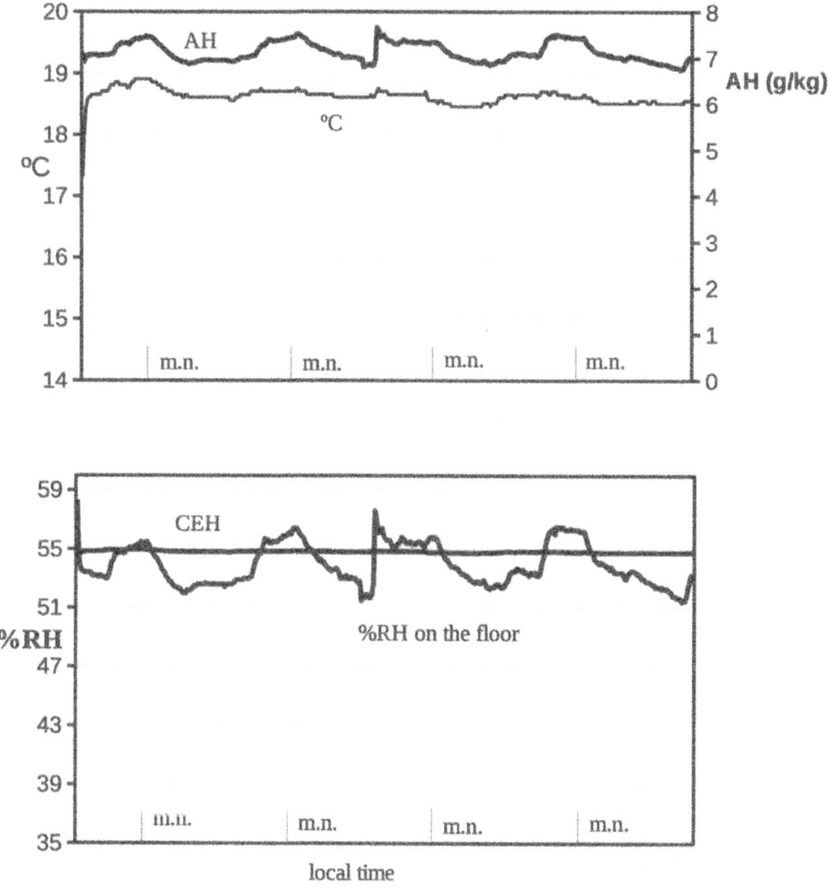

Fig. 6.4 Temperature and humidity on the ground floor of a private house in Purmerend (The Netherlands) during a four-day period (1–5 March 1999). The RH rises above the CEH for a few hours, starting in the late afternoon and lasting until after midnight. m.n., midnight. *(Unpublished data. For details of methodology, see: de Boer, R., 2003. The effect of subfloor heating on house-dust-mite populations on floors and in furniture. Exp. Appl. Acarol. 29, 315–330.)*

limited. Ventilation is usually intensified during the night. Fig. 6.5. shows an example of another period, also in winter, but the weather conditions are different now. The absolute humidity, AH, of the air is lower. As a consequence, the RH never rises above the CEH. Mites living on the floor are continuously losing body water during this period. *D. pteronyssinus* can survive this for at least 10 weeks (Section 4.3), while *D. farinae* much longer (Section 4.5).

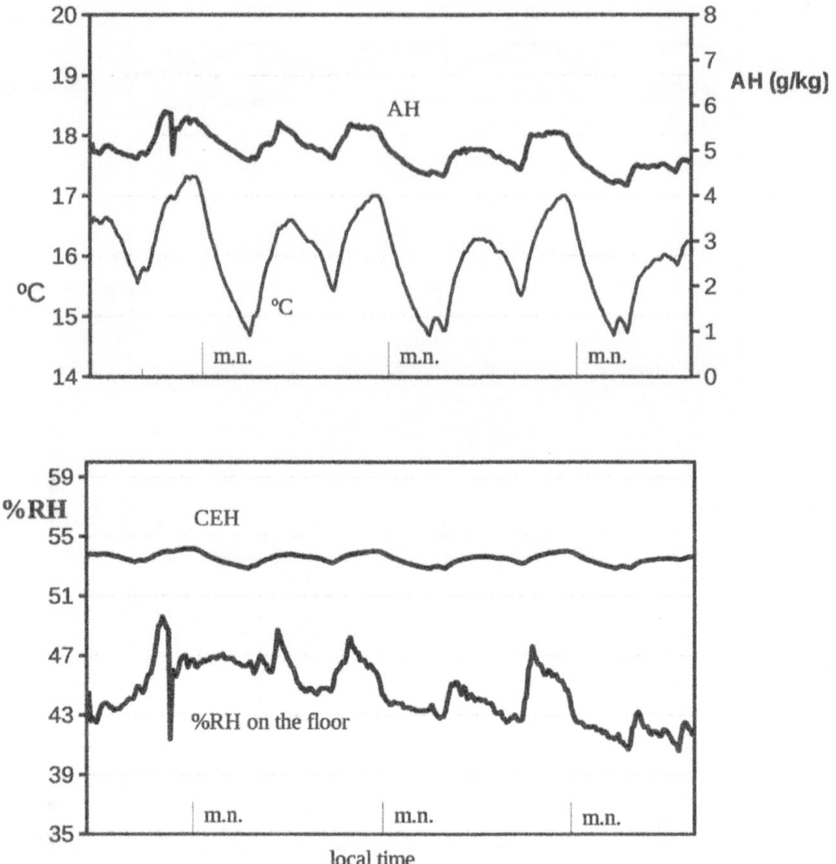

Fig. 6.5 Temperature and humidity on the ground floor of a private house in Pur-merend (The Netherlands) during a 4–day period (8–11 February, 1999) Note that the absolute humidity (AH) of the air is lower than in Fig. 6.4. The RH remains below the CEH all the time so that mites are unable to gain body water. m.n., midnight. *(Unpublished data. For details of methodology, see: de Boer, R., 2003. The effect of subfloor heating on house-dust-mite populations on floors and in furniture. Exp. Appl. Acarol. 29, 315–330.)*

During the summer, when the heating is off and the absolute humidity of the air is relatively high, RH on the floor is above the CEH more often than during the winter, but not all the time. Also during the summer, periods occur when there is insufficient water vapor in the air and dust mites cannot maintain water balance (de Boer, 1997). If these conditions persist long enough, all mites will die. But it seems that in Holland this happens rarely or never at all.

In March the winter is in retreat in Holland. In March 1994 I had the opportunity to sample the live mite population of the old, wall-to-wall carpet on the ground floor of a small dwelling in Haarlem (the Netherlands). I used the heat escape method, so only the mobile stages of live mites were captured. A total of 710 mites were captured from 14 carpet fragments from the living room, which is equivalent to 8385 mites/m^2. This massive survival of the winter is no miracle. More difficult to explain was the contrasting scarcity of mites in the corridor of the same house that was furnished with the same and equally old carpet (de Boer and Kuller, 1995 and Appendix A).

6.3 Lethal heat and cold treatments

For anyone who learns that the nasty episodes of shortness of breath are caused by house dust mites, a first reaction is likely to be: "kill them." Whether killing dust mites will indeed bring you relief from asthma symptoms is very uncertain. It depends on many more factors. Nevertheless, quite a few studies focus on killing mites.

Every organism can be killed by exposure to a high temperature. Kinnaird (1974) found that a temperature of at least 51°C is needed to kill all *D. pteronyssinus* mites in 6 h. The minimum exposure time that is required to kill 100% of the individuals is called LT-100. Chang et al. (1998), working with *D. farinae* found that at 40°C the LT-100 strongly depends on the air humidity. But at temperatures above 50°C all mites were dead within 40 min whereas the RH did not seem to play a role. Brandt and Arlian (1976) exposed *D. farinae* to a constant 34°C, 31°C, 28°C, and 25°C combined with 40 or 50%RH. The LT-100 ranged from 4 to 11 days. But 34°C did not seem to be unhealthy at all when the relative humidity was 75%. As many as 95%–97% of the mites survived these conditions for at least a week. In contrast, a slightly higher temperature (35°C) combined with 75%RH was found to be unfavorable for *D. pteronyssinus,* in a study by Koekkoek and van Bronswijk (1972). The experimental setup was such that other effects besides directly killing mites may have played a role. The authors started up cultures of *D. pteronyssinus* with five females and five males and kept these at 75%RH and either of six temperatures: 10°C, 15°C, 20°C, 25°C, 30°C, or 35°C. Thriving colonies developed at the four intermediate temperatures, 15°C, 20°C, 25°C, and 30°C, but not at 10°C and not at 35°.

Mason et al. (1999) tumble dried eight duvets for one hour. Most but not all of the resident mites were killed, whereas the temperature rose to >52°C and the RH dropped to <10%.

Tovey and Woolcock (1994) exposed two woolen carpets to the hot Australian sun during almost 6 h and found that all mobile mites appeared to be killed.

Also cold treatments can be applied to kill mites.

Paul and Sinha (1972) found that *D. farinae* adults are unlikely to survive if populations are exposed to $-18°C$ for at least 48 h. Dodin and Rak (1993) exposed *D. pteronyssinus* to $-30°C$. Humid air at this temperature killed all mobile life stages within 10 min but eggs tolerated these conditions for up to 30 min. Dry air at this temperature killed eggs more effectively, namely in less than 1 min, but some larvae, nymphs, and adults survived for up to 3 min. After 5 min all were dead. The difference in egg mortality in humid versus dry air is remarkable, but the numbers of tested specimen were not very high.

van Bronswijk and Koekkoek (1972) subjected the mobile stages of five species of pyroglyphid mites to low temperatures, namely $-1°C$, $-5°C$, $-9°C$, and $-15°C$. Lower temperatures went with shorter survival times. Some of the *D. farinae* and *D. pteronyssinus* mites survived one to several weeks at $-1°C$. At $-15°C$ all mites were dead within 6 h. Differences between the five species were rather big. It would be interesting to examine whether differences concerning cold tolerance also exist between strains of the same species. And would selection for cold tolerance result in a more cold-resistant strain?

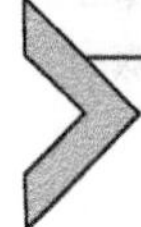

6.4 Mite control through reduction of RH, air-conditioning, mechanical ventilation, dehumidifiers, electric blankets, subfloor heating

It may be not very difficult to create desiccating conditions for dust mites in a home. The hard thing is to maintain these conditions uninterruptedly, long enough to exterminate every last mite. That may require keeping the windows tightly closed for weeks on end, which may be found unacceptable by the inhabitants. But the local climate and the season strongly influence the severity of the interventions that are required. You are lucky if you live in a place where the situation is not completely hopeless. In a warm climate it can be even desirable to keep your home tightly closed in order to have full benefit of the air-conditioner. Lintner and Brame (1993) analyzed 536 dust samples collected from 424 homes across the United States and found a significant $(P < .001)$ association of air-conditioning with reduced mite allergen levels. In contrast to air-conditioners, evaporative coolers add moisture to the air and are used especially in dry climates. Ellingson et al. (1995) examined the effect of such coolers on mite allergens in homes in Colorado (USA). They found that 12 of 19 homes with evaporative coolers were positive for mite allergen, whereas only five of 19 control homes were so (Fisher's exact test: $P = .049$, two-sided). Improved ventilation can make a difference also. Harving et al. (1991) reported that dust mites disappeared from mattresses in 11 of 16 Danish homes after installation of a mechanical ventilation system. Two studies have been done to examine the usefulness of dehumidifiers for mite control, one in England and one on the Canary Islands. The results were very different. Custovic et al. (1995) determined the mite allergen concentrations in six English houses, each one supplied with a single centrally placed portable dehumidifier and compared them with six control houses. Dust samples were taken in winter, before and 1, 2, and 3 months after the installment. The effect on the allergen concentrations turned out to be negligible. The other study was done on the Canary Islands. These islands are close to the tropics and winters are very mild, certainly in comparison with the English winters. Ten homes on these islands were enrolled in a study by Cabrera et al. (1995). In each home one room was furnished with a dehumidifier, the other one served as control. Occupants were told to use the dehumidifier at its maximum potency and keep the bedroom completely closed during the daytime whereas at night the device was programmed to maintain RH at 50%. In the morning, for approximately 20 min, the bedroom was cleaned and ventilated. Dust samples were taken in February 1993, after the dehumidifier had been operating for ≥ 3 months. Allergen of *D. farinae* (Der f1) was generally scarce and not significantly different in bedrooms with or without a dehumidifier. In contrast, levels of two other mite allergens in the two groups of rooms turned out to be significantly different when compared by the Mann-Whitney U test, which is a conservative appraisal of the findings because this test ignores the magnitude of the differences. And the magnitude of the differences was quite impressive (Table 6.2).

The temperature in a mattress can be raised by using an electric blanket. At higher temperatures house dust mites require a higher RH. The RH itself decreases with increasing temperature. Thus detrimental physical conditions for dust mites can be created by just raising the temperature. To examine the effect of this we placed electric blankets on 14 beds. By night the beds were normally slept on whereas by day the electric blankets were on for 7.25 h (de Boer and van der Geest, 1990). After 10 weeks, eight of the mattresses were dissected and the numbers of mites in fragments of the polyurethane foam

Table 6.2 Allergen concentrations in mattresses from 10 homes on the Canary Islands.

| | Dehumidifier use | | Dehumidifier use | |
Home no.	No Der p I (μg/g)	Yes Der p I (μg/g)	No Der p II (μg/g)	Yes Der p II (μg/g)
1	105.7	19.1	45	4.8
2	79.76	54.37	79	26
3	139.4	11.6	100	3.7
4	39.23	6.81	37	3.6
5	7.84	8.81	6.3	1.7
6	4.34	2.1	5.4	0.7
7	114.7	11.9	88	3.4
8	51.3	4.3	27	4.1
9	42.5	18.2	40	8.7
10	41.05	1.42	9.7	1.2
	$P = .0142^*$		$P = .0008^*$	

*Mann-Whitney U test, two-sided.
In each home, one room was supplied with a dehumidifier and one was not.
From Cabrera, P., Julià-Serda, G., Rodriguez de Castro, F., Caminero, J., Barber, D., Carrillo, T., 1995. Reduction of house dust mite allergens after dehumidifier use. J. Allergy Clin. Immunol. 95, 635–636.

core and in the ticking were determined using the heat escape method. Mites disappeared to a depth of 3.5–5.0 cm away from the electric blanket. At this distance from the electric blanket the temperature after 7.25 h is in the order of 31–45°C, whereas the RH ranged between 26% and 58%. Whether the mites disappeared because they died in the vicinity of the electric blanket or because they migrated away from it was examined in a separate experiment (de Boer, 1990). It could be shown that at least many of the mites escape by moving away from the heat source. Whether this was a directional migration or a chaotic, random movement remains an unanswered question. Yet another experiment showed that mites did not migrate back toward the electric blanket in the periods when the heating is off (de Boer, 1996).

It is also conceivable that unfavorable conditions for dust mites exist in winter on floors that are provided with subfloor heating. The same principle applies: raising the temperature causes a drop of the RH whereas mites need a higher RH. I studied 37 houses in a town near Amsterdam (de Boer, 2003). All houses had, in the living room on the ground floor, a smooth type of flooring, furnished with a rug that was at least 3 years old. Sixteen of these floors had subfloor heating. The floors with subfloor heating had, on average, fewer mites, but the difference with unheated floors was small. Only in spring was the difference statistically significant (Mann-Whitney U test, two-sided, $.02 < P < .05$). Rather high mite numbers were also found in some of the houses with subfloor heating. In fact, the house with the highest mite numbers was one with subfloor heating. This indicates that this type of heating is compatible with a thriving mite population. But perhaps the population thrives only in summer, when the heating is off. In spring, i.e., by the end of the heating season, live mites seem to be relatively scarce (but not absent) on floors with subfloor heating. It cannot be told whether these were true survivors or recent immigrants from beds, stuffed furniture, and unheated floors.

It was remarkable that mite numbers were also lower in the upholstered furniture of the houses with subfloor heating (Mann–Whitney U test, two-sided, $P < .01$). This observation was also made in other studies (Brown, 1996; Schata et al., 1990). We can only speculate about an explanation.

In each one of the houses, temperature and RH on the floor inside the rug were made every 15 min during periods lasting 4–8 days. The records show that, both in houses with and in houses without subfloor heating, the RH on floors can be below the CEH continuously for several days. At other times, undoubtedly due to weather changes, the RH in the rug was continuously above the CEH or only for a few hours every day. These periods with relatively high RH were usually in the late afternoon or early evening when household activities like cooking or bathing are going on. A moderate increase of temperature, a moderate decrease of (absolute) humidity, or a combination of both would have been enough to keep the humidity below the CEH all the time. Therefore it is conceivable that in a different climate, subfloor heating would have had a stronger effect on mite populations.

Both *D. farinae* and *D. pteronyssinus* were common and together dominated the house dust fauna. We know that *D. farinae* but not *D. pteronyssinus* can survive many months at continuously low RH (<CEH). Indeed, the houses with subfloor heating harbored relatively more *D. farinae*, but the difference is small and variable and on some floors with subfloor heating *D. pteronyssinus* and not *D. farinae* dominated.

References

Arlian, L.G., Bernstein, I.L., Gallagher, J.S., 1982. The prevalence of house dust mites, *Dermatophagoides* spp, and associated environmental conditions in homes in Ohio. Allergy Clin. Immunol. 69, 527–532.

Auer, H., Demetz, H., Frank, A., Janetschek, H., 1985. Untersuchung über das Vorkommen der Hausstaubmilbe in Tal- und Hochlagen Tirols als Voraussetzung einer Allegenkarenz-Empehlung (in German). Allergologie 8, 123–127.

Brandt, R.L., Arlian, L.G., 1976. Mortality of house dust mites, *Dermatophagoides farinae* and *D. pteronyssinus*, exposed to dehydrating conditions or selected pesticides. J. Med. Entomol. 13, 327–331.

Brown, S.K., 1996. Assessment and control of volatile organic compounds and house dust mites in Australian buildings. In: 13th Int. Clean Air & Environment Conference, 22–5 September 1996, Adelaide.

Cabrera, P., Julià-Serda, G., Rodriguez de Castro, F., Caminero, J., Barber, D., Carrillo, T., 1995. Reduction of house dust mite allergens after dehumidifier use. J. Allergy Clin. Immunol. 95, 635–636.

Chang, J.C.S., Arlian, L.G., Dippold, J.S., Rapp, C.M., Vyszenski-Moher, D., 1998. Survival of house dust mite, *Dermatophagoides farinae*, at high temperatures. Indoor Air, 34–38.

Custovic, A., Taggart, S.C.O., Kennaugh, J.H., Woodcock, A., 1995. Portable dehumidifiers in the control of house dust mites and mite allergens. Clin. Exp. Allergy 25, 312–316.

de Boer, R., 1990. Effect of heat treatments on house dust mites *Dermatophagoides pteronyssinus* and *D. farinae* in a mattress-like polyurethane foam block. Exp. Appl. Acarol. 9, 131–136.

de Boer, R., 1996. Movements of house dust mites in response to changing physical circumstances. Proc. Exp. Appl. Entomol. N.E.V. Amsterdam 7, 247–248.

de Boer, R., 1997. Explaining house dust mite infestations on the basis of temperature and air humidity measurements. In: Siebers, R., Cunningham, M., Fitzharris, P., Crane, J. (Eds.), Mites, Asthma and Domestic Design III, Wellington N.Z, pp. 13–19.

de Boer, R., 2003. The effect of subfloor heating on house-dust-mite populations on floors and in furniture. Exp. Appl. Acarol. 29, 315–330.

de Boer, R., Kuller, K., 1995. Winter survival of house dust mites (*Dermatophagoides* spp.) on the ground floor of Dutch houses. Proc. Exp. Appl. Entomol. N.E.V. Amsterdam 6, 47–52.

de Boer, R., van der Geest, L.P.S., 1990. House dust mite (Pyroglyphidae) populations in mattresses and their control by electric blankets. Exp. Appl. Acarol. 9, 113–122.

Dodin, A., Rak, H., 1993. Influence of low temperature ($-30°$C) on the different stages of the human allergy mite *Dermatophagoides pteronyssinus*. J. Med. Entomol. 30, 810–811.

Ellingson, A.R., LeDoux, R.A., Vedanthan, P.K., Weber, R.W., 1995. The prevalence of Dermatophagoides mite allergen in Colorado homes utilizing central evaporative coolers. J. Allergy Clin. Immunol. 96, 473–479.

Harving, H., Hansen, L.G., Korsgaard, J., Nielsen, P.A., Olsen, O.F., Romer, J., Svendsen, U.G., Osterballe, O., 1991. House dust mite allergy and anti-mite measures in the indoor environment. Allergy 46 (Suppl. 11), 33–38.

Hughes, A.M., Maunsell, K., 1973. A study of a population of house dust mite in its natural environment. Clin. Allergy 3, 127–131.

Kinnaird, C.H., 1974. Thermal death point of *Dermatophagoides pteronyssinus* (Trouessart, 1897) (Astigmata, Pyroglyphidae), the house dust mite. Acarologia 16, 340–342.

Koekkoek, H.H.M., van Bronswijk, J.E.M.H., 1972. Temperature requirements of a house dust mite Dermatophagoides pteronyssinus compared with the climate in different habitats of houses. Ent. Exp. Appl. 15, 438–442.

Korsgaard, J., 1983. House dust mites and absolute indoor humidity. Allergy 38, 85–92.

Lintner, T.J., Brame, K.A., 1993. The effect of season, climate, and air-conditioning on the prevalence of Dermatophagoides mite allergens in household dust. J. Allergy Clin. Immunol. 91, 862–867.

Mason, K., Riley, G., Siebers, R., Crane, J., Fitzharris, P., 1999. Hot tumble drying and mite survival in duvets. J. Allergy Clin. Immunol. 104, 499–500.

Paul, T.C., Sinha, R.N., 1972. Low-temperature survival of *Dermatophagoides farinae*. Environ. Entomol. 1, 547–549.

Schata, M., Elixmann, J.H., Jorde, W., 1990. Evidence of heating system in controlling house-dust mites and moulds in the indoor environment. In: Proc. 5th Int. Conf. Indoor Air Quality and Climate-Indoor Air '90, vol. 4, pp. 577–581.

Tovey, E.R., Woolcock, A.J., 1994. Direct exposure of carpets to sunlight can kill all mites. J. Allergy Clin. Immunol. 93, 1072–1074.

Turos, M., 1979. Mites in house dust in the Stockholm area. Allergy 34, 11–18.

van Bronswijk, J.E.M.H., 1973. *Dermatophagoides pteronyssinus* (Trouessart, 1897) in mattresses and floor dust in a temperate climate. J. Med. Entomol. 10, 63–70.

van Bronswijk, J.E.M.H., Koekkoek, H.H.M., 1972. Effects of low temperatures on the survival of house-dust mites of the family Pyroglyphidae. Neth. J. Zool. 22, 207–212.

Wickman, M., Nordvall, L., Pershagen, G., Korsgaard, J., Johansen, N., 1993. Sensitization to domestic mites in a cold temperate region. Am. Rev. Respir. Dis. 148, 58–62.

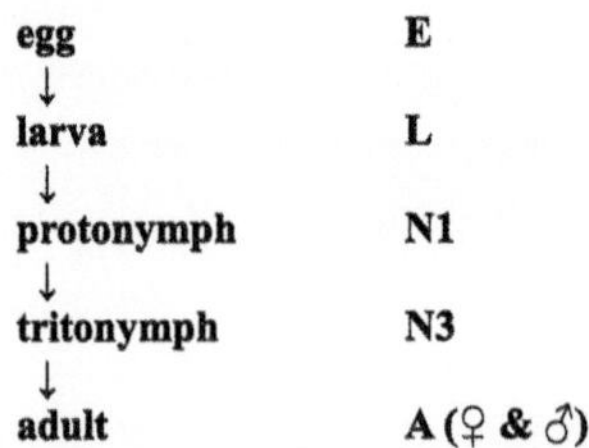

Fig. 7.1 Successive stages of the life cycle of *Dermatophagoides* spec.

CHAPTER SEVEN

The life cycle of astigmatic mites and their natural habitats

7.1 Hypopi

If you look at Fig. 7.1. you may notice something strange in the life cycle of a house dust mite. An egg gives rise to a larva and the larva gives rise to the first nymphal stage called the protonymph, N1. So far nothing unusual. But then this protonymph is followed by a tritonymph, N3 and not N2. Why did acarologists denote this stage as the third nymphal instar instead of the second?

The reason is that in many members of the Acariformes the protonymph may give rise to an instar that looks very abnormal and is totally different from the preceding and succeeding instars both in appearance and in behavior. This heteromorphic deutonymph, N2, is called a hypopus (plural: hypopi or hypopodes). And it is succeeded by a third nymphal instar, the tritonymph, N3. Hypopi lack mouthparts, have a closed gut, and are resistant to desiccation. These are adaptations to survive periods of adverse environmental circumstances. There are two kinds: active hypopi and inert hypopi. In species that form an active hypopus (Fig. 7.2.) the hypopi move freely about and are equipped with ventral suckers or claspers—adaptations to secure attachment to an insect that may carry the mite to a distant food source, which the mite could never reach by itself. This way of dispersal is called phoresy. It is a smart adaptation, because the flies and other insects are likely to have an affinity for the same patchily distributed habitats as the mites, for instance mushrooms or cadavers. And, unlike the mites, flies are equipped with wings and sense organs that enable them to locate and reach new and faraway resources.

House Dust Mites
https://doi.org/10.1016/B978-0-443-19111-4.00012-5

In contrast to the active hypopi, inert hypopi (Fig. 7.3.) are completely devoid of movement and often remain within the cuticle of the preceding, protonymphal stage. They are adapted to survive adverse conditions and may be dispersed by the wind or by animals through random contamination.

In species that form hypopi, the intercalation of a hypopus in the life cycle is facultative. It may be skipped and then the protonymph molts directly into a tritonymph. The formation of a hypopus is triggered by environmental circumstances.

In *Dermatophagoides*, however, hypopi never occur; the protonymph always molts directly into a tritonymph. So, this section of a book about house dust mites is devoted to a phenomenon that does not occur in house dust mites at all. You may find that inappropriate, but in that case I disagree. The absence of a character can be as remarkable and as intriguing as the presence of it. Let us contemplate on the formation of hypopi a little more. Why was, in the evolution of the acariform mites, the second instar chosen for hibernation and dispersal? Why the second instar? It could be argued that it makes much more sense if, instead of the immature deutonymph, the inseminated female would specialize on dispersal. Only an inseminated female is capable of founding a colony all by herself when she reaches a new, yet unoccupied habitat, for instance a bird nest. If a male or an immature mite or an unmated female would reach a vacant but suitable habitat, it would have to wait for a mate to arrive before it could start to reproduce. And it is quite conceivable that this will never happen, so that the entire perilous migration adventure is in vain. In contrast, an inseminated female usually has enough sperm stored in her body to lay scores of fertilized eggs and the sons and daughters arising from these eggs can mate among one another and produce a new generation and so on. Bingo!

So, if inseminated females were equipped with adaptations for dispersal, we would have no problem to give a satisfactory explanation for it. But now we have to explain why adult females are not equipped with such adaptations, whereas in subadult, virgin females, such adaptations are well developed.

Nature seems to abhor self-fertilization. By avoiding self-fertilization it is avoided that mutant, defective genes combine in the homozygous state. The first statement ("nature seems to abhor self-fertilization"; perhaps the most frequently quoted one-liner from the writings of Charles Darwin) is an observation. The second statement (avoidance of self-fertilization serves to avoid deleterious effects of recessive genes) is merely an interpretation. It seeks to explain the observation that nature seems to abhor self-fertilization. If this explanation is right, not only self-fertilization but also less severe forms of inbreeding like sister-brother matings or niece-nephew matings should be abhorred by nature. Perhaps here we have an explanation for the evolution of dispersal by deutonymphs rather than by adult females. Perhaps avoidance of inbreeding is the impetus behind it.

As discussed later (Section 7.3), inbreeding is also relevant in connection with the sex ratio. A surplus of females can be explained as a consequence of inbreeding. A 50:50 sex ratio, i.e. equal numbers of males and females, is associated with "pan mixes," i.e. with the absence of inbreeding. And *D. pteronyssinus* has a 50:50 sex ratio, which may indicate a *panmictic* population structure.

It is noteworthy that in *Dermatophagoides farinae*, it is the quiescent protonymphal stage preceding the formation of a pharate tritonymph that may be prolonged to help the mite survive long periods of dehydrating circumstances, as discussed in Section 4.6. Thus in this mite species the stage that is adapted to survive adverse circumstances is intercalated in the life cycle at the point, where in other astigmatic mites the heteromorphic deutonymph is formed. And like the heteromorphic deutonymph the intercalation of it is facultative. But there are differences. There is no extra molt. Morphologically, the prolongated quiescent protonymph is similar to a normal quiescent protonymph. It is not adapted for dispersal. Like a normal quiescent protonymph, the ventral side of it is glued to the substrate.

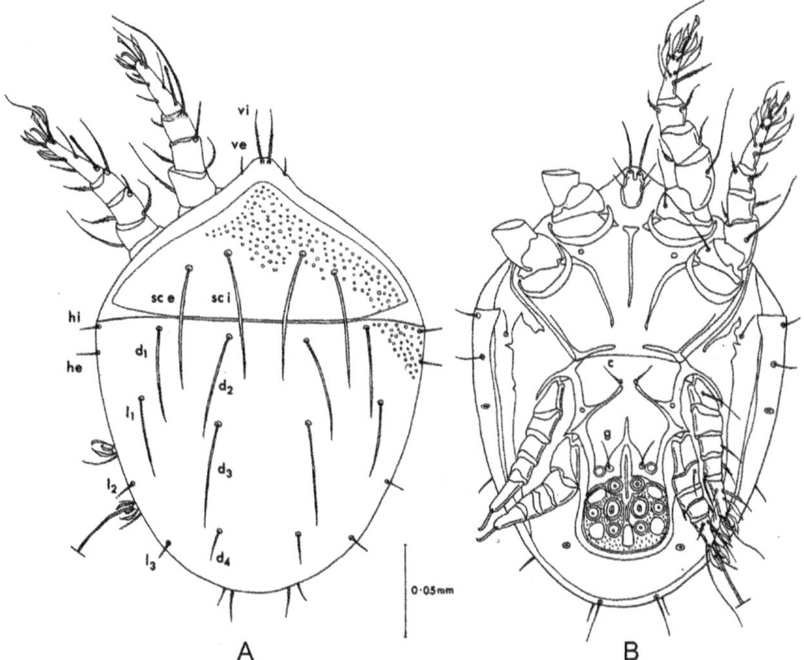

Fig. 7.2 Hypopus of *Acarus siro*—example of an active hypopus. (A) Dorsal view, (B) ventral view. *(From Hughes, A.M., 1976. Mites of Stored Food and Houses: Technical Bulletin 9. London: Ministry of Agriculture, Fisheries and Food. ISBN: 0-11-240909-1.)*

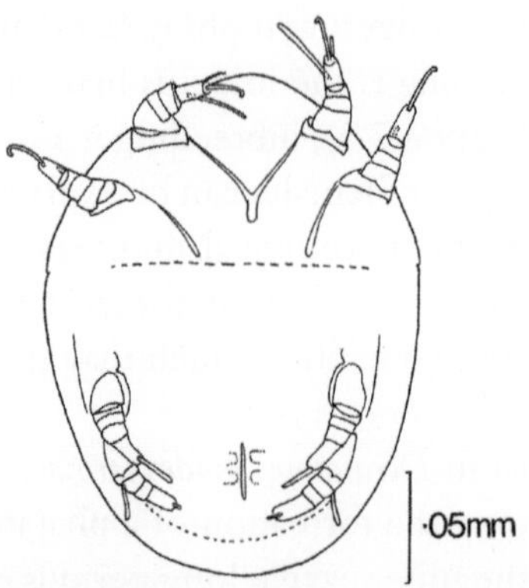

Fig. 7.3 Hypopus of *Lepidoglyphus destructor*—example of an inert hypopus. Ventral view. *(From Hughes, A.M., 1976. Mites of Stored Food and Houses: Technical Bulletin 9. London: Ministry of Agriculture, Fisheries and Food. ISBN: 0-11-240909-1.)*

7.2 The natural habitats and colonization

Dust mites belong to the mite family *Pyroglyphidae*. Most of the known species were originally discovered in the nests of birds. Fain et al. (1990) and Colloff (2009) made a compilation of data concerning the species of bird or mammal on which, or in the nest of which, pyroglyphid mites were found. I tried to find some information on the breeding behavior of the animals involved (Appendix E). A total of 69 mite-host associations, involving 42 different mite species, were recorded. Only three records concern an association between a pyroglyphid mite and a feral mammal, invariably a rodent. All the other feral associations concern birds as hosts for the mites. This apparent preponderance of birds does not necessarily reflect the situation existing in nature. Bird nests are much easier to find and to collect than mammal nests. Those mammal species that do make a nest often have their nest in a burrow deep underground. Underground nests of mice are all around us but very difficult to find. They are hiding acarological secrets. So the importance of mammal nests as habitats for pyroglyphid mites may be underestimated. But there is no doubt about the importance of bird nests.

I have been interested in birds since I was a kid and that brought me some knowledge of bird nesting behavior. With this knowledge in mind, I find it hard to imagine how a mite species could adapt successfully to a habitat like a bird nest. Many passerine birds make a free nest in a bush or a tree and then often a new nest is constructed every year. Nests in tree holes provide a more enduring resort for mite populations. But how can mites get there? For a pyroglyphid mite, tree holes are like tiny islands, sparsely spread in an immense ocean that is otherwise uninhabitable, in fact deadly hostile. The only way to get from one nest to another one is by adhering to the feathers of a visiting bird. But that is only good if that bird flies from one tree hole to another one, which usually they don't. Birds return to their own nest repeatedly to incubate or feed the nestlings, but they will not visit the nest of another bird couple. Only at the start of the breeding season hole nesting birds will fly from one tree hole to another one in order to select the most suitable one. Perrins (1980) in his book entitled *British Tits* writes (p. 154–155): "Since the female often breeds in the site in which she roosts it seems that she is usually responsible for making the final decision over the choice of nesting sites, possibly some months before breeding. In all species the pairs explore together, the male showing sites to the female. … The male commonly shows his mate several sites before she accepts one."

So it does happen that a bird flies from one tree hole to another one, but only rarely. Tree holes are also used by birds to sleep in. But that will not involve frequent movements from one hole to another one. The chances that a mite will reach another tree hole by adhering to the feathers of a visiting bird must be slight. If I were a mite living in a tree hole, I would stay there and never take the risk of leaving. Leaving would be a crazy thing to do. On the other hand, every mite living in a tree hole is the descendant of a mite that once took the risk of leaving. The innate tendency to attach to the feathers of a visiting bird must be favored by selection in spite of the risks involved.

In Appendix E, I categorized the bird species according to breeding habits, in particular the utilization of tree holes or crevices for building a nest. A bird nest built in a tree hole or a crevice would appear to be more suitable for mites than a 'free' nest between the branches of a bush or a tree. Yet, among the 41 recorded bird hosts of pyroglyphid mites 13 concern bird species that normally build a free nest. Another six recorded bird hosts I classified as birds breeding gregariously. Perhaps gregarious breeding enhances the exchange of nidicoles. Also gregarious roosting by one of the partners of a pair of birds may perhaps explain the presence of nidicolous mites in their nest.

Of the 42 mite species recorded by Fain et al. (1990) and Colloff (2009), 12 were recovered from house dust in human dwellings only. The common house dust mites, *D. pteronyssinus*, *D. farinae*, and *Euroglyphus maynei*, were not recorded from any bird nest or any nest of a feral mammal. Perhaps the investigators were primarily interested in taxonomy and ignored the more common species. Anyway, there is a surprising contrast with the findings in a more recent investigation, namely by Silva et al. (2018). That was primarily a faunistic, ecological study. It was done in Brazil. Fifty-two nests of birds were sampled using a Berlese funnel, which means that the mites were captured alive. A total of 76 mite species were recorded. Some were ectoparasites, some were predators, some were accidental stragglers, and some may have been more or less adapted to a life in a bird nest. The common dust mites *E. maynei* and *D. farinae* were absent, but there were quite a few *D. pteronyssinus*. The nests of 10 bird species, out of a total of 26, yielded at least a few specimens of *D. pteronyssinus*. Some nests had so many that the mites must have been reproducing there. For example, a nest of the wren, *Troglodytes musculus*, yielded 963 specimens of this mite. Unfortunately, the authors give no details about that nest. Some species of wren are known to build their nest in a cavity or a bird house. The habit of puncturing the eggshell of birds nesting in their territory has been described for the house wren. Transportation of mites could be a side effect of this behavior.

Two nests of Great Kiskadees (*Pitangus sulphuratus*) yielded 418 *D. pteronyssinus*. Wikipedia gives the following information about the breeding behavior of the Great Kiskadee: "Both sexes build a large domed nest that has a side entrance. It is chiefly composed of grasses and small twigs but can also incorporate lichen, string and plastic. The birds will steal material from other nests. The nest is placed in a wide range of sites, often in an exposed position high up in a tree or on man-made structures. Occasionally the nest is placed in a cavity."

A nest of the white-rumped swallow (*Tachycineta leucorrhoa*) yielded 130 specimens of *D. pteronyssinus*. This swallow builds nests in holes or crevices in trees. It will sometimes reuse favorable nest sites and it occasionally breeds in abandoned nests of the firewood gatherer, a bird in the family *Furnariidae*. The white-rumped swallow is known to display nest prospecting behavior, visiting potential future nesting sites. Nest prospecting is a behavior recorded in both breeding and nonbreeding individuals (Wikipedia).

For other birds it is more difficult to imagine how the mites reached their nests. For instance, a total of 14 specimens of *D. pteronyssinus* were found in the nests of three species of doves (*Columbidae*). Doves make a free nest of twigs in a bush or a tree. I could not suggest a satisfactory explanation for how these mites may have arrived there.

7.3 Dispersal of domestic house dust mites

Mollet and Robinson (1995) performed a unique experiment to demonstrate the ease by which house dust mites are dispersed through a dwelling or from one house to another one. They used a laboratory colony of *D. farinae*. The mites were marked with the dye Sudan Red 7B, a stain that is taken up into the hemocoel and the tissues, where it disappears only slowly. Two experiments were done. In the first one, approximately 1850 marked mites were released onto a couch on the downstairs level of a two-level residence. The mites were left undisturbed overnight. Nine hours later two small children were placed on the couch for about 3 h and their clothes were subsequently examined. Fifty-one marked mites could be recovered from the clothes. Ten days after releasing the mites, the upholstery, furnishings, and carpet of the residence were vacuum sampled in various places. Eight marked mites were recovered, all upstairs and in the carpet. A second trial was done to determine if mites would use clothing to disperse to the family vehicle. Approximately 3750 red mites were released onto the downstairs couch. After 9 h two children were placed on the couch for about 3 h, after which they sat in the vehicle for approximately 1 h. Twenty-four hours later the vehicle was vacuum sampled. One red mite was recovered. Unfortunately, no information was given about the life stages (nymph or adult) of the introduced and the recovered mites.

When sticky tape is stuck to the upper side of a mite-infested carpet fragment, mites will incidentally get stuck on the tape. Bischoff et al. (1992) showed that the number of mites caught on the tape can be enhanced by warming the fragment on a heat plate. Mites move more and faster at higher temperatures, so it is no miracle that more mites get trapped, even if they move randomly in all directions. Brown (1994) applied a heat source (a bottle filled with warm water) to the upper side of mite-infested material supplied with sticky tape on the surface. He found indications, albeit not very strong, that mites move away from the heat source. If that is really so, it would mean that mites can sense the location of a heat source and migrate, directionally, away from it. It causes one to wonder whether mites will also move directionally toward or away from a person, sitting on an infested couch—a behavior that would have implications for mite dispersal. Perhaps the direction of the movements, away from or toward the heat source, depends on the temperature—a possibility that is amenable to experimental verification.

The studies were done using laboratory cultures. The mites were subjected for generations to totally different conditions than the free-ranging dust mites that we wish to understand and that we wish to control. Selection pressure concerning migration behavior properties must have been very different. This probably changed the innate behavior of these mites. It is good to keep that in mind.

7.4 The sex ratio

The primary sex ratio of *Dermatophagoides pteronyssinus* is 1:1, i.e. equal numbers of males and females. It could be argued that this is biologically illogical because a greater number of females than males would favor the increase rate of populations.

Suppose every female mite would have in her progeny one son for every 10 daughters. One son is quite enough to fertilize 10 females. And a son cannot himself lay eggs, as daughters can. So it is good for the survival of the species to have more daughters than sons. A newly founded population would increase its size much faster than with equal numbers of males and females. The validity of this explanation would be easily accepted *if* populations had many more females than males. But the real world is different. Sexually reproducing organisms usually have as many males as females or almost as many. The explanation that is given to us by theoretical biologists runs as follows:

Imagine a hypothetical population where the sex ratio is *not* 1:1. Imagine, for instance, a population in which 90% of the individuals are female. Females of this population give birth to nine daughters for every son. In a population of 100 individuals there would be 90 females and 10 males. If you are a female in such a population, 1/90 of all individuals in the next generation will be your offspring. That is only 1.1%. All other individuals, nearly 99%, are the progeny of other females. But if you are a male 1/10 of all individuals of the next generation is your progeny. That is 10%. Ten times more than if you were a female. So an average male has a much greater contribution to the next generation than an average female.

That is an unstable situation in an evolutionary sense. Mutations occur all the time and some of them affect the sex ratio. Suppose the effect of one mutation is that a female gets more sons and fewer daughters; for instance, 20% sons instead of only 10%. That female will have more grandchildren than the other females. Many of these grandchildren carry the mutant gene that causes females to have more sons. As a consequence, the frequency of the mutant gene will increase. And as a consequence of that, the sex ratio of the population will shift toward more males.

More mutations will eventually lead to a population with as many males as females. Then the process will stop. An equilibrium is reached. A stable equilibrium in an evolutionary sense. It is a good example of what is known as an Evolutionary Stable Strategy, or more briefly: an ESS (Maynard Smith, 1978).

It must be emphasized that this is so only in a so-called panmictic population where the progenies of the females are thoroughly mixed in every new generation. When, in contrast, there is some degree of inbreeding, i.e. when close relative matings such as sister-brother or niece-nephew matings are more common, then a 1:1 sex ratio is not an ESS. That is easy to see if you imagine a population where sister-brother matings are the rule, i.e. where all the daughters of a female are inseminated by her own sons. The best she can do in this situation is to give birth to just enough sons to have all her daughters inseminated. In many organisms one son will be able to inseminate all his sisters. Having more sons would be a waste of time and resources. An extremely female-biased population would evolve. In an intermediate situation, when inbreeding is common but not so extreme, a less extremely female-biased sex ratio is expected to evolve (see Charnov, 1982).

Whether you believe or don't believe that evolution really works this way is, of course, your own choice. But the facts are that we generally see a 1:1 sex ratio in sexually reproducing organisms in spite of the fact that males cannot give birth (by definition). A female-biased sex ratio indicates that the occurrence of matings between close relatives in the population is increased. Conversely, a 1:1 sex ratio is evidence of random mating. And dust mites do have a 1:1 sex ratio. In populations of dust mites and nidicolous mites, random mating depends on frequent and intensive dispersal. So it can be inferred that a 1:1 sex ratio is also evidence of frequent and intensive dispersal. It must be possible to obtain observational evidence for this corollary.

7.5 Males, females, copulation, and egg production

When, in the evolution of the arthropods, the transition was made from life in the sea to life on land, breathing air was only one of the changes the organisms had to deal with. Controlling evaporation, i.e. controlling loss of body water, was another one. Fertilization, i.e. bringing together and uniting egg cells with sperm cells, was yet another problem. Arthropods that live in the sea simply discharge their sperm in the surrounding water. A sperm cell swims to a conspecific egg cell and merges with it. On the land that is not possible. The problem has been solved by the evolution of internal fertilization. Fusion of gametes takes place inside the female body. For us, humans, so accustomed to internal fertilization, it may be strange to think of it as a novelty. But for the conquest of the land by arthropods and other taxa the evolution of internal fertilization was critical. Among members of the mite suborder *Oribatida* sperm transfer is usually indirect, mediated by a stalked spermatophore. In the spermatophore, sperm is packaged and protected against desiccation. It is deposited somewhere by the male. Females actively search for spermatophores and internalize it when they find one. However, dust mites and other members of the cohort *Astigmatina* depart from the general oribatid pattern. Male dust mites do not produce spermatophores. They engage in a true copulation. Sperm is transferred to the female directly through a special opening called the "bursa copulatrix." And fertilized eggs leave the female body through another, separate opening, the ovipositor (Figs. 7.4–7.17).

Thus male and female dust mites have to come together. How they find each other is not known. Trying to find a mate by random walk in a sparsely inhabited carpet must be tedious. Possibly smell helps to locate a partner from a distance. Smelly volatiles, technically known as pheromones, are perhaps released by the opisthosomal glands (see Section 7.5.1).

Males are equipped with suckers which they use to attach to the female during copulation. Two big suckers are near the anus and smaller ones are on the tarsi of the fourth pair of legs. Not only adult females have the attention of the males. Also females that are still in the pharate stage, prior to the final ecdysis, are attractive for males. When observed through a dissecting microscope males can be seen waiting near a pharate mite. This behavior is seen in other mite species as well and is referred to as "guarding." To a human observer a pharate female dust mite looks the same as a pharate male, but a male house dust mite somehow senses a difference. A pharate mite that is guarded by a male invariably molts into a female. How the males distinguish between the sexes is not known, but again pheromones may be involved. Another observation that hints at the involvement of pheromones was made by Alexander et al. (2002). They found that *D. farinae* females live on for quite a long time after they reached their maximum egg production and stopped laying. And during this postreproductive period they are no longer attractive for males.

The penis of the male is pointed forward when in rest. When the male is ready for mating the penis turns backward and the copulatory sac is protruded.

Spieksma observed copulations of *D. pteronyssinus* through a dissecting microscope and gives the following eyewitness account in his thesis (Spieksma, 1967): "The male attaches itself by the two adanal suckers to the posterior part of the back of the female such that their gnathosomata (plural of gnathosoma) point in opposite directions. The fourth pair of legs of the male grasps the female laterally. The third pair makes regular movements along the dorsal part of the lateral side of the female. The male may remain attached to the female for several hours. In these laboratory observations the moment of copulation could not be determined exactly."

Guarding a female serves not only to secure the first opportunity to mate. After mating the male may go on guarding the female to prevent other males from inseminating her a second time and insert competing sperm. Adult females can be seen walking about with a male on their back.

Humans and other mammals do not store sperm. Their females have to be (re)inseminated over and over again, at least once for every new litter. Insects and mites are more efficient. Females of these organisms have an organ called the "receptaculum seminis," in which sperm can be stored. One copulation may be sufficient to fill the receptaculum seminis with enough sperm to fertilize all the eggs that she is going to lay in her entire life. However, Spieksma (1967) observed that, after the production of 25–50 eggs by a female *D. pteronyssinus*, a second insemination is needed to continue egg laying. Even a third insemination may eventually be required to go on. Also for *D. farinae* it has been found that a second insemination may be needed to continue egg production (Alexander et al., 2002).

Arlian et al. (1990) give detailed information about the life history of 56 isolated, adult females at 23°C, 75%RH and with unlimited access to food. After insemination egg laying soon commences, but not sooner than 24 h after emergence. Usually the first egg is laid after 3–4 days. Egg production during subsequent days is shown graphically in Fig. 7.18. Death occurred on average 1.8 days (range 1–8) after cessation of egg production. It was also shown that uninseminated, i.e. non-egg-laying females live much longer than egg-laying females, namely 77.4 days (s.d.: 34.5, range: 19–140), as compared with 31.2 days (s.d.: 11.1) for inseminated, egg-laying females. Also Matsumoto et al. (1986) found, both for *D. pteronyssinus* and *D. farinae*, that unmated females live longer than mated ones and that this effect of mating status does not occur or not so strongly in males. So it seems that egg production is exhausting and that refraining from laying eggs improves longevity. Theoretically that has important ecological consequences. Egg production under less favorable and more realistic circumstances is less well documented. Undoubtedly the daily egg production will be much less when the temperature is low or food is scarce. When food is scarce it takes more energy and more time to find enough food to produce eggs. But ultimately the total egg production is not necessarily less. Possibly oviposition will continue longer, thus compensating for a slower egg production. It would be interesting to examine whether starving females live longer than females who have access to food. A female who fails to find a mate necessarily postpones reproduction. Again, her reproductive output may be partly or perhaps even fully compensated for by an ongoing egg production during a longer life. It would be interesting to examine how many adult but uninseminated females are present in a carpet. How many virgin females walk around in your carpet, obsessed with the burning desire to meet a conspecific of the opposite sex? This could be found out by making use of the "Hirschmann method" (Section 2.2) to capture females alive and keep them isolated to see if they are able to lay eggs. It may tell a tale about the structure and dynamics of dust mite populations. Suppose that uninseminated, adult females are very scarce or absent. That would mean that emerging adults quickly find a mate. Not unlikely because we know that dust mites in carpets have a clustered distribution. But suppose that many mites leave their birth ground, by adhering to the clothes of people, before they reach maturity. These are likely to end up in a place where mites are scarce and, if they survive, they are destined to remain virgin for a long time. But they have avoided insemination by a close relative.

7.5.1 Sense organs and pheromones

Insects and mites can communicate by smell, releasing volatile chemicals that are recognized by their conspecifics. These volatiles are referred to as semiochemicals or pheromones.

The common, cosmopolitan, astigmatic mite, *Tyrophagus putrescentiae* (cheese mite or mold mite), showed evidence of emitting an alarm pheromone when it was disturbed. An extract of mite bodies in pentane was prepared and analyzed by Kuwahara et al. (1975). The constituents of the extract were separated by chromatography in a silicic acid column. Fractions were subjected to a simple bioassay to test their repellent effect on live mites: An impregnated piece of filter paper is placed on a glass plate that is crowded with mites. If the filter paper contains a repellent volatile the mites move away from it, leaving a deserted area around it. The fraction with the strongest repellent effect was further analyzed and a compound named "neryl formate" ($C_{11}H_{18}O_2$) emerged as the alarm pheromone.

Emitted volatiles were also trapped using a Porapak Q column and subsequently analyzed by gas-liquid chromatography. If the mites emitting the volatiles were either disturbed or undisturbed, the chromatograms looked totally different. Also differences in the color of the latero-abdominal glands were noted between mounted specimens of disturbed and undisturbed mites. This could be interpreted as an indication that these glands produce the alarm pheromones (Kuwahara et al., 1979). Observations on other mite species similarly provided evidence of the involvement of the latero-abdominal glands in the production of pheromones (Kuwahara et al., 1980).

Also *D. pteronyssinus* and *D. farinae* emit the volatile compound neryl formate; however, in these species it does not function as an alarm pheromone but as an aggregation pheromone. This was shown by Skelton et al. (2010) using a Y-tube olfactometer: A mite is placed at the base of the main leg of the Y-shaped tube where it either starts walking or does not start walking, against a gentle air stream that is carrying a volatile. When the mite reaches the junction in the Y-tube, it has to make a choice. Either it chooses for the arm that furnishes the volatile or it chooses the other arm that supplies clean air. Male and female *D. pteronyssinus* and *D. farinae* chose significantly more often ($P < .05$) for the arm that brought air enriched with neryl formate. Using a Y-tube olfactometer it could also be shown that *D. farinae* mites are attracted by food (fish flakes) volatiles (Skelton et al., 2007).

Y-tubes are not suitable to demonstrate the effect of close-contact pheromones. A fraction of a hexane extract of *D. farinae* containing the compound HMB (2-hydroxy-6-methylbenzaldehyde) did not elicit a reaction of *D. pteronyssinus* or *D. farinae* in a Y-tube experiment. However, HMB stimulated mounting behavior, i.e. copulation attempts, when *D. farinae* was

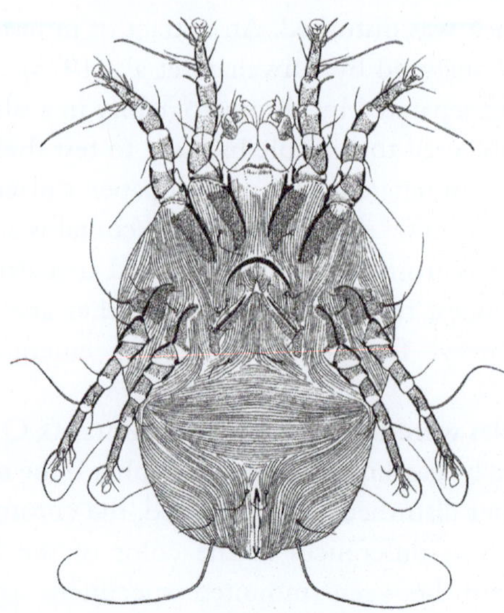

Fig. 7.4 Female *Dermatophagoides pteronyssinus*, ventral view. *(From Fain, A., 1966. Nouvelle description de* Dermatophagoides pteronyssinus *(Trouessart, 1897) Importance de cet acarien en pathologie humaine (Psoroptidae: Sarcoptiformes). Acarologia 8, 302–327.)*

Fig. 7.5 Male (left) and female *D. pteronyssinus* in copula. *(Photography by Bert Mans.)*

exposed to pieces of filter paper impregnated with this compound (Tatami et al., 2001).

So dust mites must possess chemoreceptors. Evidence of photoreceptors was presented by Furumizo (1975). He observed that *D. farinae* responds differently when exposed to light of different wavelengths. So they seem to have photoreceptors but where these are is not known.

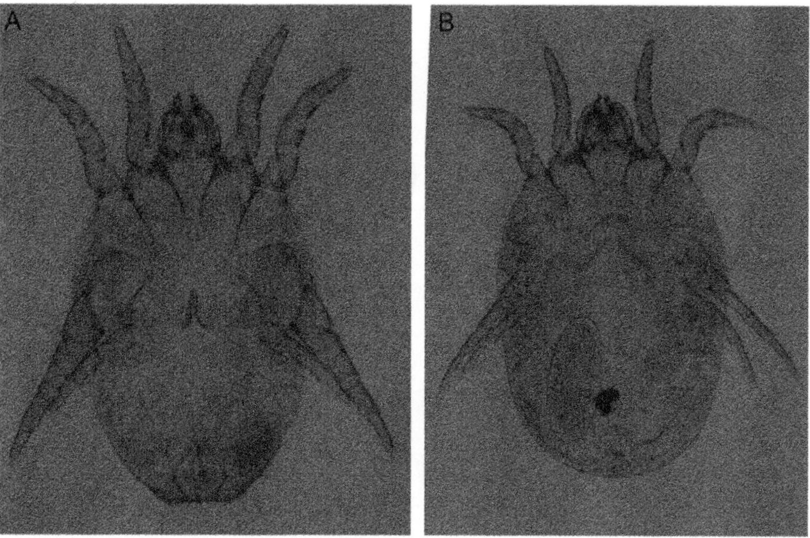

Fig. 7.6 *Dermatophagoides pteronyssinus*. Left: male. Note the penis, anal suckers, and the size of the third pair of legs. Right: female. Note the epigynum, vulva, and the egg inside her body. *(Photography by Kees Kuller.)*

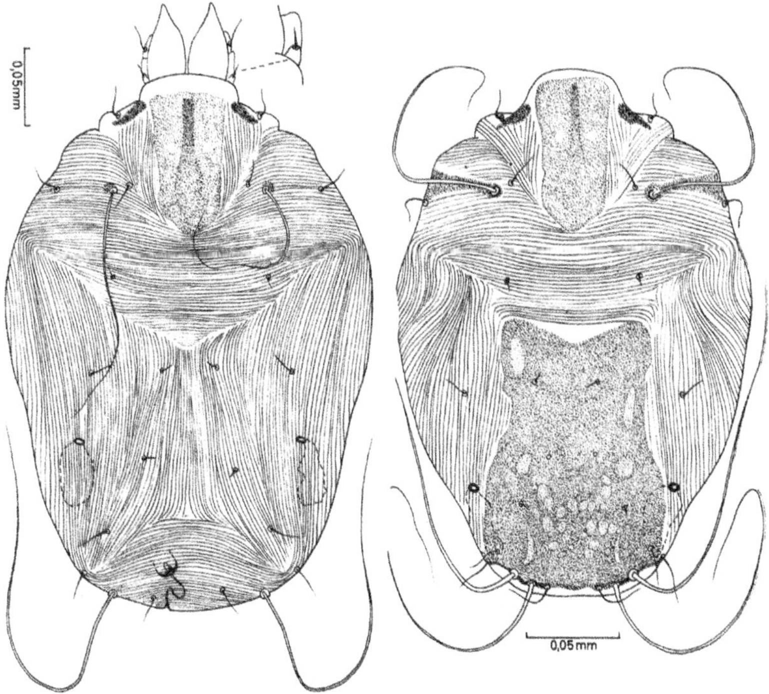

Fig. 7.7 Dorsal view of female (left) and male (right) *Dermatophagoides pteronyssinus*. A male has a hysterosomal shield (stippled posterior half of the dorsal side), while a female does not. *(From Fain, A., 1966. Nouvelle description de Dermatophagoides pteronyssinus (Trouessart, 1897) Importance de cet acarien en pathologie humaine (Psoroptidae: Sarcoptiformes). Acarologia 8, 302–327.)*

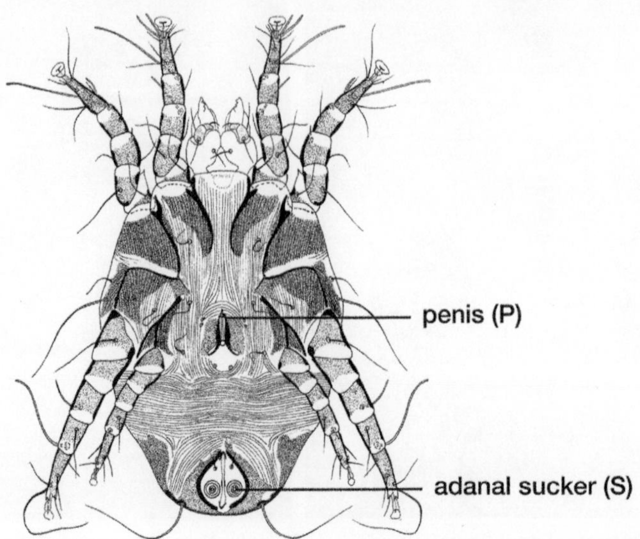

Fig. 7.8 Sclerotized parts of the genital organs of a male *D. pteronyssinus*. Drawing based on cleared specimen prepared for light microscopy. Ventral side. *(From Fain, A., 1966. Nouvelle description de Dermatophagoides pteronyssinus (Trouessart, 1897) Importance de cet acarien en pathologie humaine (Psoroptidae: Sarcoptiformes). Acarologia 8, 302–327.)*

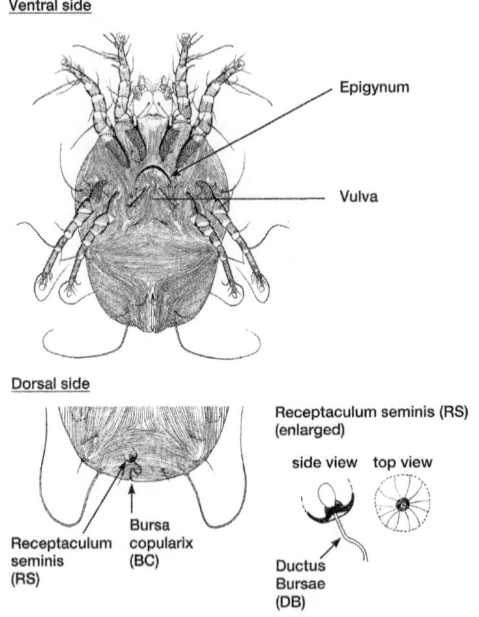

Fig. 7.9 Sclerotized parts of the genital organs of a female *D. pteronyssinus*. Drawing based on cleared specimen prepared for light microscopy. The bursa copulatrix is the site for introduction of sperm. Epigynum and vulva are the sclerotized parts associated with the ovipositor, the organ for egg deposition. *(Modified after Fain, A., 1966. Nouvelle description de Dermatophagoides pteronyssinus (Trouessart, 1897) Importance de cet acarien en pathologie humaine (Psoroptidae: Sarcoptiformes). Acarologia 8, 302–327.)*

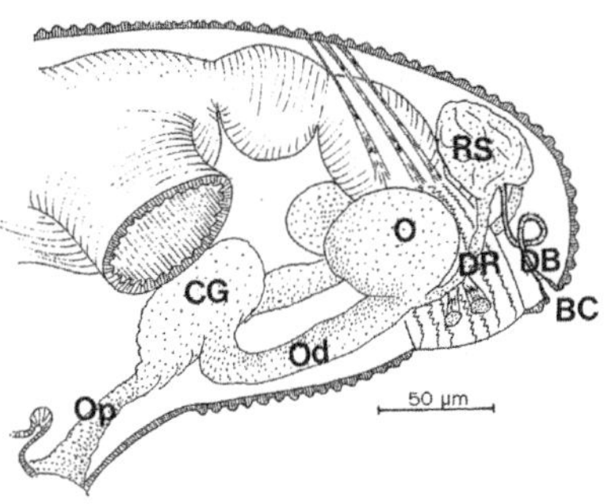

Fig. 7.10 Schematic drawing of the genital organs of a female *D. pteronyssinus.*

Symbol	Name	Function
BC	Bursa copulatrix	Insertion of aedeagus (penis) of the male and introduction of sperm
DB	Ductus bursae	transport of sperm from BC to RS
RS	Receptaculum seminis	Storage of sperm
DR	Ductus receptaculi	Transport of sperm from RS to O
O	Ovary	Formation of egg cells
Od	Oviduct	Transport of egg cells from O to CG
CG	Chorion gland	Finishing the (fertilized) egg, including the formation of the chorion (egg shell)
Op	Ovipositor	Deposition of the egg

(From Walzl, M.G., 1992. Ultrastructure of the reproductive system of the house dust mites Dermatophagoides farinae *and* D. pteronyssinus *(Acari, Pyroglyphidae) with special remarks on spermatogenesis and oogenesis. Exp. Appl. Acarol. 16, 85–116.)*

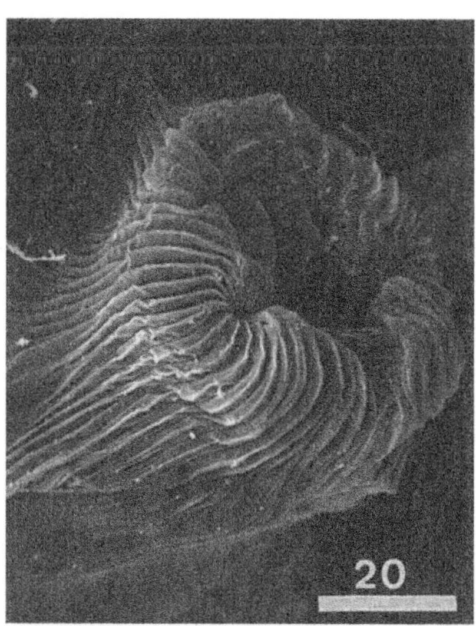

Fig. 7.11 SEM of ovipository organ of *D. farinae* ready for oviposition, ventral view. Scale bar in μm. *(From Walzl, M.G., 1991. Comparison of the chitinous parts of the reproductive organs of house dust mites by means of scanning electron microscopy. In: Schuster, R., Murphy, P.W., (Ed.), The Acari; Reproduction, Development and Life-History Strategies. Chapman & Hall, London, pp. 355–362.)*

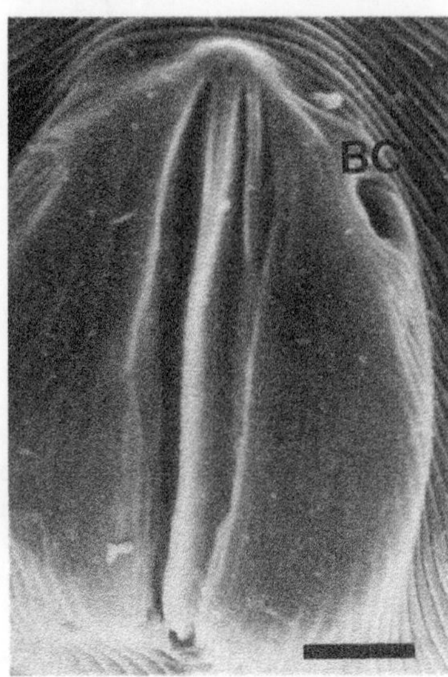

Fig. 7.12 Anus and bursa copulatrix (BC) of female *D. pteronyssinus*. Scale = 10 µm. *(From Walzl, M.G., 1992. Ultrastructure of the reproductive system of the house dust mites* Dermatophagoides farinae *and* D. pteronyssinus *(Acari, Pyroglyphidae) with special remarks on spermatogenesis and oogenesis. Exp. Appl. Acarol. 16, 85–116.)*

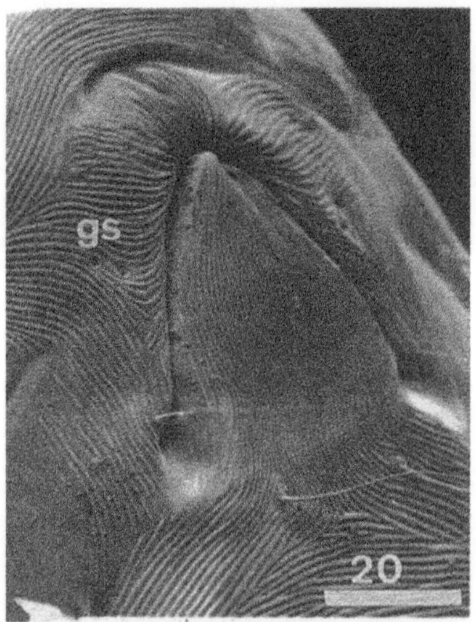

Fig. 7.13 SEM of ovipository organ of *D. farinae* in inoperative position, ventral view, gs: genital "sucker." Scale bar in µm. *(From Walzl, M.G., 1991. Comparison of the chitinous parts of the reproductive organs of house dust mites by means of scanning electron microscopy. In: Schuster, R., Murphy, P.W. (Ed.), The Acari; Reproduction, Development and Life-History Strategies. Chapman & Hall, London, pp. 355–362.)*

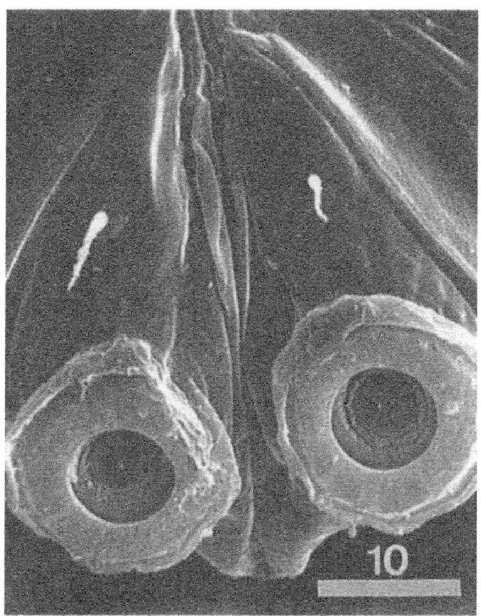

Fig. 7.14 Anal region of *D. farinae* showing the protruded anal suckers. Scale bar in μm. *(From Walzl, M.G., 1991. Comparison of the chitinous parts of the reproductive organs of house dust mites by means of scanning electron microscopy. In: Schuster, R., Murphy, P.W. (Eds.), The Acari; Reproduction, Development and Life-History Strategies, Chapman & Hall, London, pp. 355–362.)*

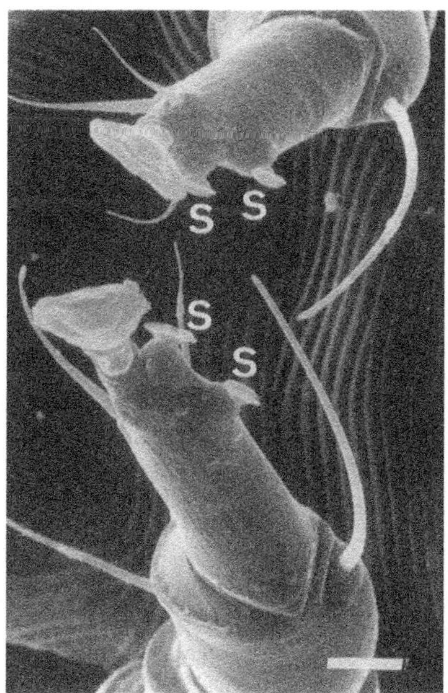

Fig. 7.15 Suckers (S) on the tarsi on the fourth pair of legs of *D. pteronyssinus*. Scale =5 μm. *(SEM picture from Walzl, M.G., 1992 Ultrastructure of the reproductive system of the house dust mites Dermatophagoides farinae and D. pteronyssinus (Acari, Pyroglyphidae) with special remarks on spermatogenesis and oogenesis. Exp. Appl. Acarol. 16, 85–116.)*

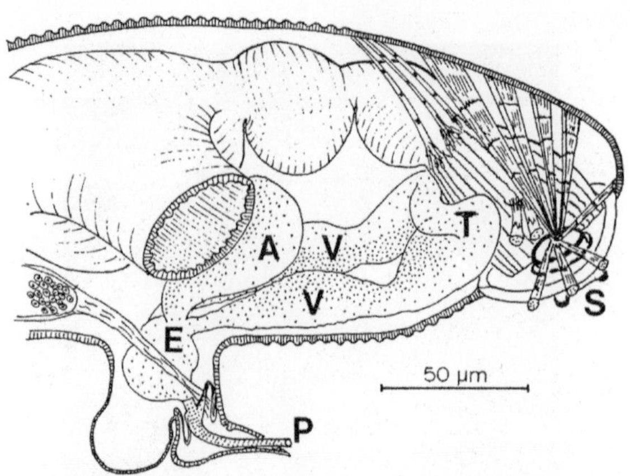

Fig. 7.16 Schematic drawing of the genital organs of a male *D. pteronyssinus.*

Symbol	Name	Function
S	Adanal suckers	Attachment to female in copulation
T	Testis	Meiosis; formation of sperm cells
V	Vas deferens (paired)	Transport of sperm cells from T to E
A	Accessory gland	Formation of sperm fluid
E	Ejaculatory duct	Transport of sperm to P
P	Penis	Transfer of sperm to female

(From Walzl, M.G. 1992 Ultrastructure of the reproductive system of the house dust mites Dermatophagoides farinae *and* D. pteronyssinus *(Acari, Pyroglyphidae) with special remarks on spermatogenesis and oogenesis. Exp. Appl. Acarol. 16, 85–116.)*

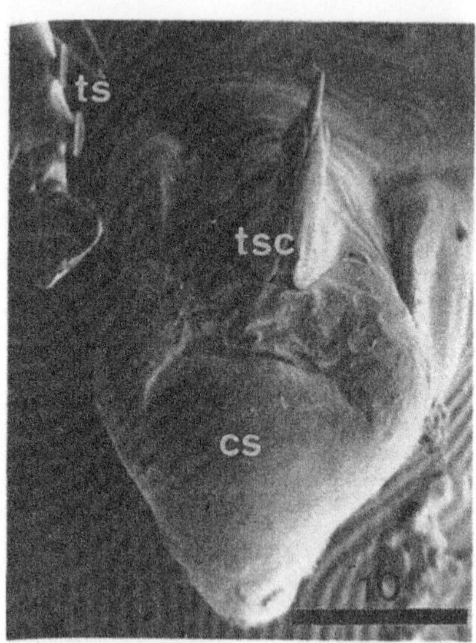

Fig. 7.17 SEM of ventral view of protruded copulatory organ of *D. pteronyssinus.* ts = tarsal sucker; tsc = transmission sclerite; cs = copulatory sac. *(From Walzl, M.G., 1991. Comparison of the chitinous parts of the reproductive organs of house dust mites by means of scanning electron microscopy. In: Schuster, R., Murphy, P.W. (Eds.), The Acari; Reproduction, Development and Life-History Strategies. Chapman & Hall, London, pp. 355–362.)*

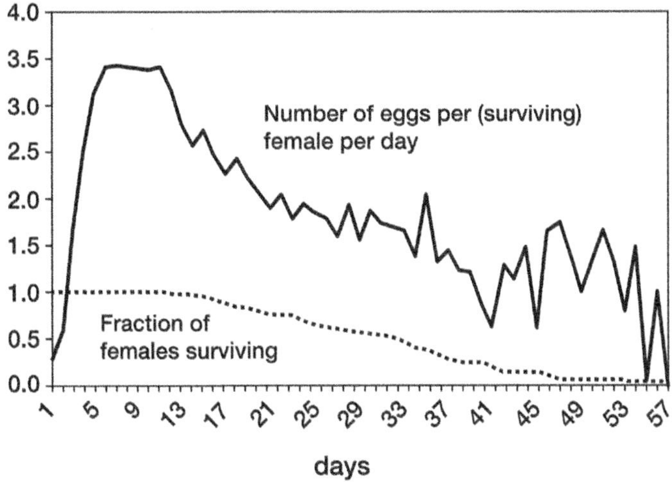

Fig. 7.18 Day-by-day egg production and survival of 56 *D. pteronyssinus* females, starting from the first day of adulthood (day 1) at 23°C, 75%RH, and food ad libitum. *(Data were read from a graph presented by Arlian, L.G., Rapp, C.M., Ahmed, S.G., 1990. Development of Dermatophagoides pteronyssinus. J. Med. Entomol. 27, 1035–1040.)*

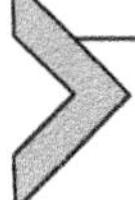

7.6 The subadult stages of *Dermatophagoides*

7.6.1 The egg stage

Sexually reproducing organisms always have a single-celled stage in their life cycle, namely the fertilized egg. That is a good starting point for a description of the ontogenetic development. In optimal conditions, a female *D. pteronyssinus* produces one to four eggs per day. In mounted specimen, an egg can sometimes be seen inside the body of a female (Fig. 7.6B). Under the most favorable physical conditions the embryonic development takes about a week. Thus almost the entire development occurs after the egg has been deposited. However, Sesay and Dobson (1972) studied the acarofauna of mattresses in Glasgow and reported the following observation: "Normally it (*D. pteronyssinus*) is oviparous but twice fully developed larvae were seen within females. This phenomenon was also noted by Bronswijk and Sinha (1971)." de Saint Georges-Gridelet (1987) reports: "Larval development inside intra-uterine eggs is observed when *D. pteronyssinus* females are submitted to various types of stress such as heat, dryness, cold, overpopulation or fungal proliferation." To explain these observations, there are at least two possibilities: Perhaps the female died and an egg in her body continued to develop. But it is also possible that eggs are retained longer in the female's body when circumstances, in particular food supply, do not allow a high egg production rate.

A peculiar feature of some acarid groups and some other arachnids as well is the occurrence of a prelarva. The prelarva (plural: prelarvae) is an immobile, nonfeeding stage that is passed inside the egg shell, prior to the development of the larva. Its origin and its function (if any) are a matter of speculation. Possibly it constitutes the rudiments of an ancestral, mobile stage that, for some reason, regressed in the evolutionary past of these organisms. In dust mites and in other *Oribatida*, the prelarva is little more than a sac without legs or mouth parts. Such extreme reduction of the prelarva has been called 'calyptostasis' (Greek: kalyptos = hidden). Fain (1977) and Fain and Herin (1978) accurately studied the prelarva of *D. pteronyssinus* by light microscopy. His interpretation of what he saw is shown in Fig. 7.19. Two sclerotized, hemispherical excrescences with pointed apices are situated in the apical third of the egg. About these structures Fain remarks: "When one examines these excrescences in lateral position it becomes clear that they are not part of the shell nor of the larva itself but that they are an intermediate structure lying between the two" and "The role of these structures is probably to break the shell of the egg and to allow the larva to escape." His observations are supplemented by those of Walzl (1988), who used scanning electron microscopy (SEM) to study the embryonic development of *D. farinae*. Walzl remarks that: "About 100 hours after oviposition two egg teeth become visible at the basis of the chelicerae, sign of the finished chitinisation of the prelarva" and "The hatching larva splits the egg shell longitudinally, leaving the prelarval integument as exuvia in the egg shell together with the egg teeth (Figs. 7.20 and 7.21)."

7.6.2 The larval and nymphal stages

It is the business of an adult mite to eat and reproduce. It is the business of a subadult mite to eat and grow. The durations of the immature stages at 23°C and 75%RH recorded by Arlian et al. (1990) are presented in Tables 7.1 and 7.2. The first mobile stage, the larva, took longer than the other immature stages. This was so also at 16°C, 30°C, and 35°C. Spieksma (1967) recorded durations of immature stages of 65 *D. pteronyssinus* mites at 25°C and 80%RH. In this strain the duration of the larval stage was not longer than the other nymphal stages.

In microscopic preparations, larvae can be easily distinguished from all other stages because they have only three pairs of legs. Three instead of four pairs of legs in the larval stage is typical for the acari. The two other preimaginal stages of *Dermatophagoides,* the protonymph and the tritonymph, can be distinguished on the basis of the genital stigmata, also called the genital papillae (Figs. 7.22 and 7.23). Larvae have no genital stigmata, protonymphs possess one pair and tritonymphs, like the adults, possess two pairs. Some authors referred to the genital stigmata as "the genital sense organs." Others suggest that "the genital papillae or genital suckers are involved in secretion of adhesive substances for copulation" (see Fain et al., 1990 p. 144). This seems unlikely because these papillae are flanking the ovipositor, the opening for depositing eggs, which is not the same opening where sperm is introduced—the bursa copulatrix. Walzl (1991) concludes from his SEM studies of *D. farinae* and *D. pteronyssinus* that "the so-called genital suckers ... represent two pairs of pits in the cuticula where muscles are attached. There is no indication that they have a sensory function."

Arthropods. You don't have to like them. But don't call them primitive. My fascination for the molting process is something that has intensified through the years. The molting process (ecdysis) marks the transition from one mobile stage to the next one. The first thing we all learned about it is that arthropods have to molt because the outer skeleton that envelops their body is rigid and cannot grow bigger so it must be replaced by a bigger one from time to time. The outer skeleton has a function in locomotion in concert with the attached muscles. That function is temporarily abandoned. A molting mite is immobile and attached to the substratum. The posture of the immobile mite inside the old cuticle is characteristically with the tips of the legs folded back. The larva, when it is still inside the egg, has the same posture, albeit with one pair of legs less (Fig. 7.21D). The protective function of the exoskeleton is also adjourned during molting. A molting mite is inevitably more vulnerable than a mobile one. But the protection against desiccation is a function of the exoskeleton that cannot be interrupted. In an early stage of the molting process, the old cuticle separates from the underlying epidermal cells (apolysis). The new integument is formed on the outside of the epidermis. So the new envelop will fit tightly around the body, whereas the old integument is still surrounding it. After the old cuticle is shed, the new one must get bigger. So it has to be flexible, at least for a while, and all the time it has to remain watertight.

Brody (1969), who did elaborate microscopic studies of *Dermatophagoides* wrote: "The laminated integument of mites is secreted by the hypodermal cells. Minute pore canals and larger ducts extend from the hypodermis, traverse the integument, and ultimately carry secretory material through the endocuticle to the surface of the exocuticle. These waxy secretions solidify as the protective epicuticle". Beament (1959) described the removal of this covering from various arthropods by abrasion, melting, and dissolution … and demonstrated the protective function of the epicuticle since dehydration of the test animals ensued.

The third ecdysis marks the transition from subadult to adult. A quiescent tritonymph develops into a pharate adult. A pharate adult guarded by a male is almost certainly a female. For an experimentalist this offers a possibility to identify and isolate females before they are inseminated, a prerequisite for doing crossing experiments.

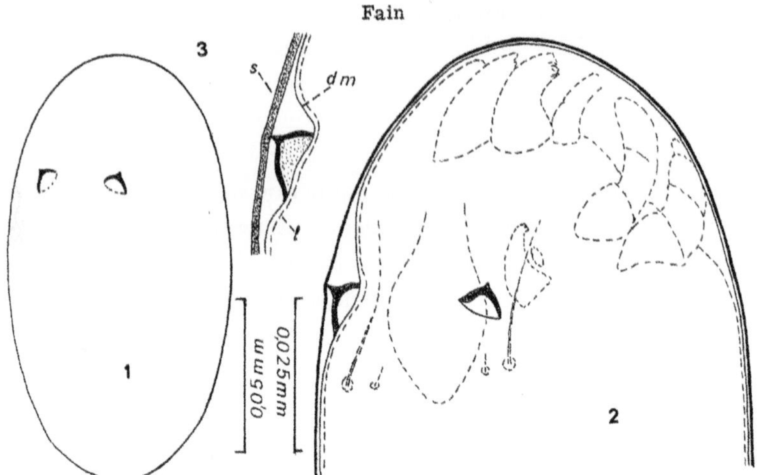

Fig. 7.19 Nr. 1. Nonlarvigerous intrauterine egg of *Dermatophagoides pteronyssinus* with sclerotized excrescences (specimens from New York, August 1968); Nr. 2. Larvigerous intrauterine egg of *D. farinae* showing the sclerotized excrescences (specimen from Singapore, XI. 1968); Nr. 3. part of Nr. 2 enlarged: *l* = larva (in broken lines), *dm* = deutovial or prelarval membrane, *s* = egg shell. *(Picture and caption from: Fain, A., 1977. The prelarva in the Pyroglyphidae (Acrina: Asti). Int. J. Acarol. 3, 115–116.)*

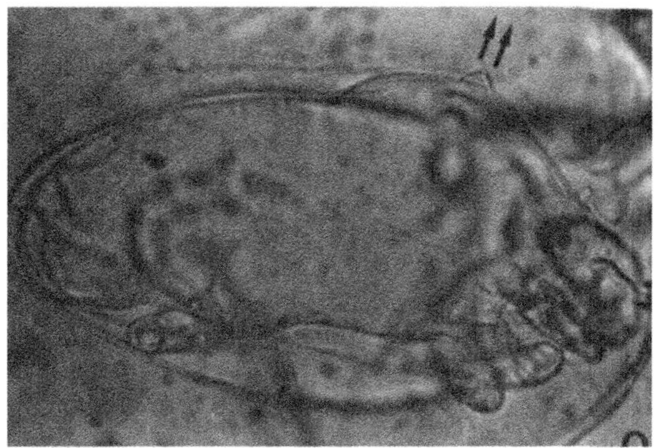

Fig. 7.20 Microscopic preparation with illustration of hatching phenomenon. Visible dehiscence organ (*arrows*). (*Picture and caption from: de Saint Georges-Gridelet, 1987.*)

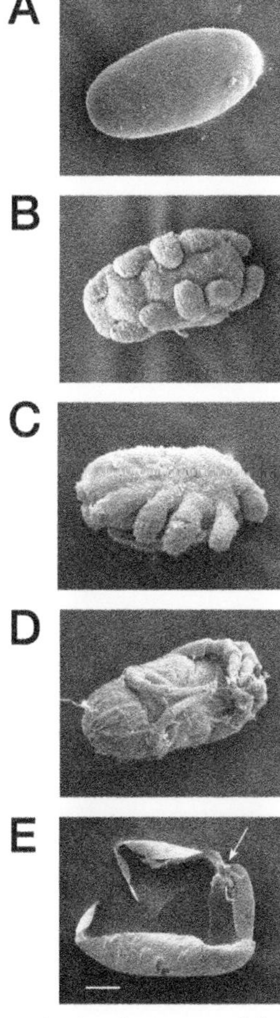

Fig. 7.21 SEM showing embryonic development of *D. farinae*. (A) Egg, (B–D) embryo inside egg (egg shell removed), (B, C) young embryo showing, from right to left, the cheliceral, pedipalpal, and three ambulatory limb buds, (D) older embryo showing the typical way in which the tips of the legs are folded back (similar as in the pharate nymphal stages), and (E) shed egg shell with egg teeth (*arrow*) and remnants of the prelarva. Scale bar: 20 µm. The micrographs were kindly provided by professor M.G. Walzl (Vienna, Austria).

Table 7.1 Duration (mean ± SD) of the immature life stages of a strain of *Dermatophagoides pteronyssinus* under near optimum physical conditions (23°C and 75%RH) and with food ad libitum.

Stage	Total duration (days)	Quiescent (days)
Egg	8.1 ± 0.9	
Larva	10.4 ± 3.2	4.0 ± 1.4
Protonymph	6.9 ± 2.0	3.2 ± 1.0
Tritonymph	8.3 ± 3.5	3.4 ± 1.5
Adult mated female	31.2 ± 11.1	
Preoviposition period	1–6	
Oviposition period	23.3 ± 10.8	
Postoviposition period	1–8	
Adult unmated female	44.7 ± 23.8	
Male	77.4 ± 34.5	

From Arlian, L.G., Rapp, C.M., Ahmed, S.G., 1990. Development of *Dermatophagoides pteronyssinus*. J. Med. Entomol. 27, 1035–1040.

Table 7.2 Developmental time from egg deposition to adult emergence (mean ± SD).

°C	Days (mean ± SD)	N	% Developed
35	15.0 ± 2.0	90	87
30	19.3 ± 2.5	137	81
23	34.0 ± 5.9	175	86
16	122.8 ± 14.5	95	59

From Arlian, L.G., Rapp, C.M., Ahmed, S.G., 1990. Development of *Dermatophagoides pteronyssinus*. J. Med. Entomol. 27, 1035–1040.

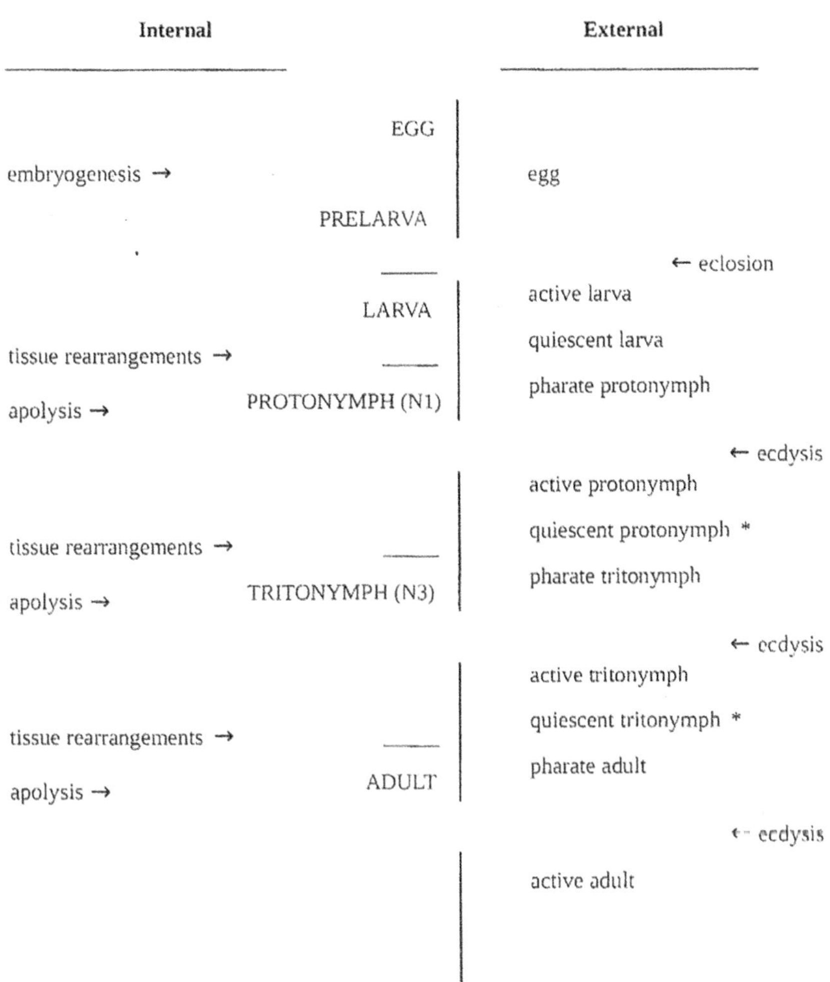

* The quiescent protonymph and also, but less often, the quiescent tritonymph stage may be much prolonged in *Dermatophagoides farinae*.

Fig. 7.22 Schematic presentation and terminology concerning the life cycle of *Dermatophagoides* spec.
Pharate is the instar within the previous cuticle prior to ecdysis (molt).
Apolysis is the separation of the old cuticle from the underlying epidermal (hypodermal) cells.
Ecdysis (molt) is the act of splitting and shedding a cuticular layer of structures.

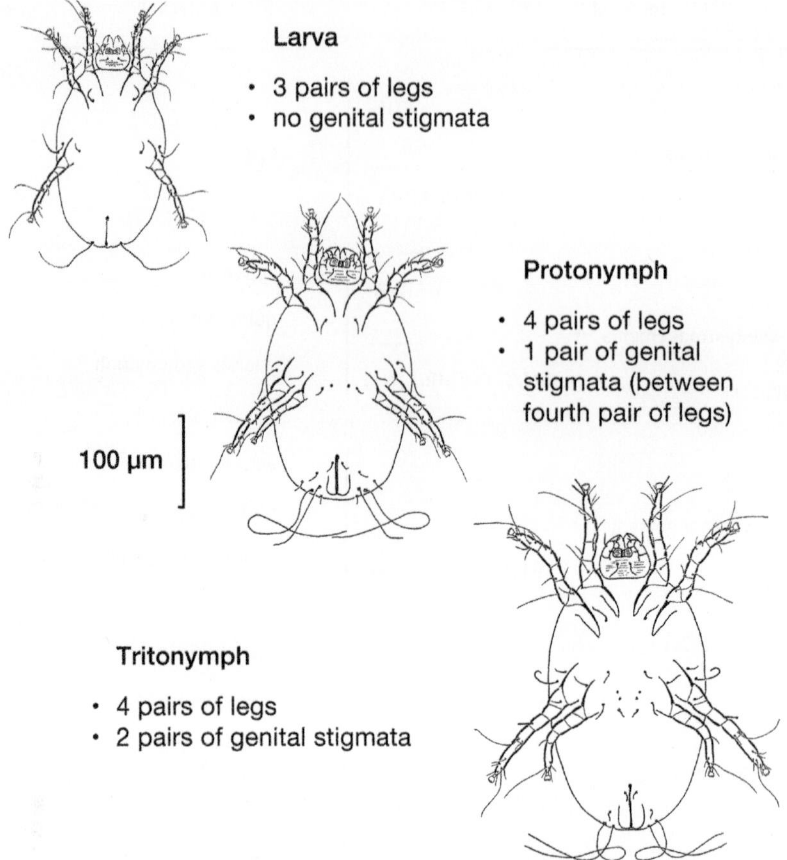

Fig. 7.23 Ventral side of the three immature stages of *D. pteronyssinus. (Artwork by the author, using drawings from Mumcuoglu, Y., 1976. House dust mites in Switzerland I. Distribution and taxonomy. J. Med. Entomol. 13, 361–373.)*

References

Alexander, A.K., Fall, N., Arlian, L.G., 2002. Mating and fecundity of Dermatophagoides farinae. Exp. Appl. Acarol. 26, 79–86.

Arlian, L.G., Rapp, C.M., Ahmed, S.G., 1990. Development of *Dermatophagoides pteronyssinus.* J. Med. Entomol. 27, 1035–1040.

Beament, J., 1959. The waterproofing mechanism of arthropods. J. Exp. Biol. 36, 391–422.

Bischoff, E., Fischer, A., Liebenberg, B., 1992. Assessment of mite numbers: new methods and results. Exp. Appl. Acarol. 16, 1–14.

Brody, A.R., 1969. Comparative fine structure of Acarine integument. J. N.Y. Entomol. Soc. 77, 105–116.

Brown, S.K., 1994. Optimisation of a screening procedure for house dust mite numbers in carpets and preliminary application to buildings. Exp. Appl. Acarol. 18, 423–434.

Charnov, E.L., 1982. The Theory of Sex Allocation. Princeton University Press.

Colloff, M.D., 2009. Dust Mites. CSIRO/Springer Science, ISBN: 978-90-481-2223-3. 583 pp.

de Saint Georges-Gridelet, D., 1987. Destruction of eggs of *Dermatophagoides pteronyssinus* (Acari: Pyroglyphidae) by natamycin and imidazoles in vitro. Int. J. Acarol. 13, 5–14.

Fain, A., 1977. The prelarva in the Pyroglyphidae (Acrina: Asti). Int. J. Acarol. 3, 115–116.

Fain, A., Herin, A., 1978. La prélarve chez les Astigmates (Acari). Acarologia 20, 566–571.

Fain, A., Guerin, B., Hart, B.J., 1990. In: Guerin, B. (Ed.), Mites and allergic disease. Allerbio, Varenne en Argonne, France. 190 pp.

Furumizo, R.T., 1975. Laboratory observations on the life history and biology of the American house dust, Dermatophagoides farinae (Acarina, Pyroglyphidae). California Vector News 3, 49–59.

Kuwahara, Y., Ishii, S., Fukami, H., 1975. Neryl formate: alarm pheromone of the cheese mite, *Tyrophagus putrescentiae*. Experientia 31, 1115–1116.

Kuwahara, Y., Fukami, H., Ishii, S., Matsumoto, K., Wada, Y., 1979. Pheromone study on acarid mites II presence of alarm pheromone in the mold mite, Tyrophagus putrescentiae (Schrank) (Acarina: Acaridae) and the site of its production. Jpn. J. Sanit. Zool. 30, 309–314.

Kuwahara, Y., Fukami, H., Ishii, S., Matsumoto, K., Wada, Y., 1980. Pheromone study on acarid mites III Citral: isolation and identification from four species of acarid mites, and its possible role. Jpn. J. Sanit. Zool. 31, 49–52.

Matsumoto, K., Okamoto, M., Wada, Y., 1986. Effect of relative humidity on life cycle of the house dust mites, *Dermatophagoides farinae* and *D. pteronyssinus*. Jpn. J. Sanit. Zool. 37, 79–90.

Maynard Smith, J., 1978. The Evolution of Sex. Cambridge University Press, ISBN: 0-521-29302-2.

Mollet, J., Robinson, W., 1995. Use of marked mites to study the dispersal of the American house dust mite (*Dermatophagoides farinae* Hughes). In: Tovey, E., Finfoot, A., Sieber, L. (Eds.), Mites Asthma and Domestic Design II. University Printing Service, Univ. Sydney, pp. 19–21.

Perrins, C.M. (Ed.), 1980. Collins, London., ISBN: 0-00 219537-2. British tits.

Sesay, H.R., Dobson, R.M., 1972. Studies on the mite fauna of house dust in Scotland with special reference to that of bedding. Acarologia 14, 384–392.

Silva, D.E., Da Silva, G.L., Moreira Do Nascimento, J., Ferla, N.J., 2018. Mite fauna associated with bird nests in Southern Brazil. Syst. Appl. Acarol. 23, 426–440.

Skelton, A.C., Birkett, M.A., Pickett, J.A., Cameron, M.M., 2007. Olfactory responses of medically and economically important mites (Acari: Epidermoptidae and Acaridae) to volatile chemicals. J. Med. Entomol. 44, 367–371.

Skelton, A.C., Cameron, M.M., Pickett, J.A., Birkett, M.A., 2010. Identification of Neryl Formate as the airborne aggregation pheromone for the American house dust mite and the European house dust mite (Acari: Epidermoptidae). J. Med. Entomol. 47, 798–804.

Spieksma, F.T.M., 1967. The House-Dust Mite Dermatophagoides Pteronyssinus (Trouessart, 1897), Producer of the House-Dust Allergen (Academic thesis). Rijks Universiteit Leiden, Leiden.

Tatami, K., Mori, N., Nishida, R., Kuwahara, Y., 2001. 2-Hydroxy-6-methylbenzaldehyde: the female sex pheromone of the house dust mite *Dermatophagoides farinae* (Astigmata, Pyroglyphidae). Med. Entomol. Zool. 52, 279–286.

Walzl, M.G., 1988. A scanning electric microscopic study of the embryonic development of the house dust mite, *Dermatophagoides farinae* (Pyroglyphidae, Actinotrichida) from Blastula to the hatching of the larva. Inst. Phys. Conf. Ser. 93, 159–160.

Walzl, M.G., 1991. Comparison of the chitinous parts of the reproductive organs of house dust mites by means of scanning electron microscopy. In: Schuster, R., Murphy, P.W. (Eds.), The Acari; Reproduction, Development and Life-History Strategies. Chapman & Hall, London, pp. 355–362.

CHAPTER EIGHT

Modeled population dynamics

Ceci n'est pas une pipe. This is not a pipe. This sentence, in big letters, is painted on a painting by the Belgian surrealist artist René Magritte (1898–1967). On the same painting above this writing, you see the image of what is unmistakably a very big pipe. You are not looking at a pipe but merely at a picture of a pipe. Whether Magritte just likes to be funny or has something more serious in mind, I don't know. For a scientist, however, there is certainly a serious message here. Scientists use mathematical models and computer simulations in order to explain the real world. What they explain is really a model of the real world, not the real world itself.

Many years ago I attended a presentation by an ecologist who himself was much dedicated to ecological field trials. His presentation was entitled "Dangers of armchair ecology." Modeling in ecological research is what he called "armchair ecology." The dangers it may entail were highlighted. I agree that we may never blindly rely on the outcomes of a model. Whatever a model predicts must be checked by observations and experiments. On the other hand, I think that the use of models in ecology is indispensable and inevitable. Whenever I am riding on my bike in my home town, as Dutch people commonly do, I always use a city plan. This city plan is in essence a model, a grossly simplified replica of the town, which I carry ... in my head. Without it I inevitably get lost in my own home town.

All I want to say is that there is nothing unusual about using models. We do it all the time. Everybody does. Not just scientists.

The city plan that I carry in my head is a very simple model—a model that my mind can manage. For more complicated models we need pen and paper or a computer. But nevertheless these are merely models. You may look at it this way: The present chapter is not really about *Dermatophagoides pteronyssinus*. It is about hypothetical organisms having properties in common with *D.p.*

House Dust Mites
https://doi.org/10.1016/B978-0-443-19111-4.00011-3

8.1 Population development in circumstances with constant optimum physical conditions and with unlimited access to food

(Mathematical details concerning this section can be found in Appendix F.)

Survival and egg production of 56 females of a strain of *Dermatophagoides pteronyssinus* under near optimum physical conditions (23°C and 75%RH) and with food ad libitum were recorded day by day by Arlian et al. (1990). Based on their data I constructed a more compressed life and fertility table, with survival and egg production figures for successive periods lasting 8 days. Eight days roughly equal the duration of the immature life stages of *D. pteronyssinus*. This way I could calculate the changes in the numbers of eggs and of mites of different life stages in subsequent periods of 8 days. I am sure that a skilled mathematician would make several refinements in order to have a closer resemblance to the real *D. pteronyssinus* in an environment that allows unlimited optimal growth. But the results would be essentially the same.

The development of a mite population arising from a single inseminated female is depicted in Fig. 8.1. As can be seen, the relative proportions of the various life stages develop toward a stable composition, with eggs > larvae > protonymphs > tritonymphs > adults. Almost half of all mites are in the egg stage.

By day 224, the age distribution is practically constant and the size of the population is increasing exponentially according to the formula:

$$N(t) = N(0)e^{rt}$$

where $N(0)$ is the initial number of mites and eggs and $N(t)$ is the number of mites and eggs at some later time t.

The parameter r is a constant number, known as "the instantaneous coefficient of population growth" and is an important parameter in ecology. It is also referred to as the "intrinsic rate of (natural) increase" or as the "Malthusian parameter." By the procedure followed for this projection, the value of r is found to be 0.0743 per head per day (i.e. 0.520 per head per week, see Appendix F1 and 2).

For the calculations of changes in the developing population structure, I assumed that a mite of less than 8 days old is an egg and a mite of between 8 and 16 days old is a larva, etc. Furthermore, I did not distinguish between active and quiescent mites. For a more accurate calculation of r and the stable age distribution based on a more detailed life table, with one-day instead of eight-day periods, the procedure described by Birch (1948) was followed. The results are presented in Table 8.1 (details in Appendix F3).

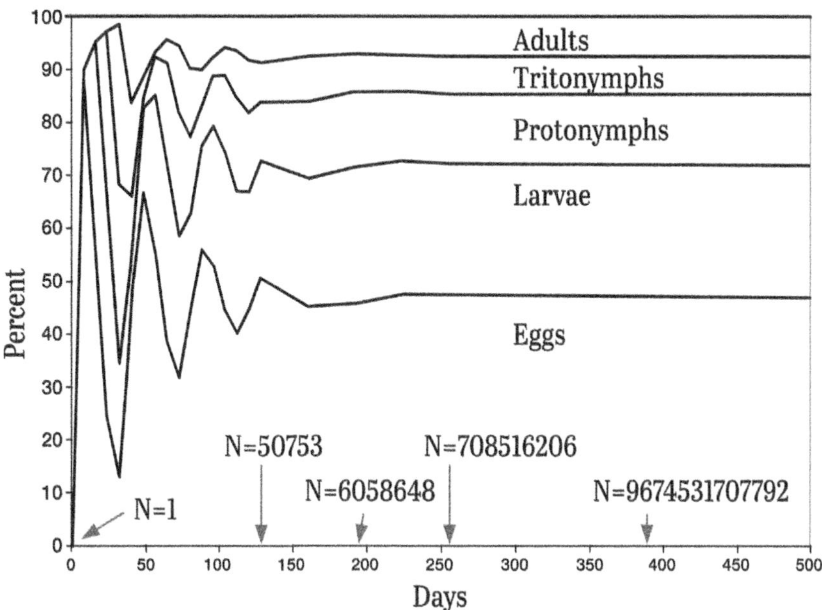

Fig. 8.1 Population development under constant optimum conditions, starting with a single inseminated female. Only female offspring is counted. N = total number of female mites and eggs.

Table 8.1 The stable stage distribution and r.

	Total	Quiescent
% Eggs	46.4	
% Larvae	29.3	9.1
% Protonymphs	10.5	3.8
% Tritonymphs	6.7	2.0
% Adults	7.1	
% Quiescent mites	14.9	
r (per head, per day)	0.0729	
Doubling time (days)	9.51	

A big, female *D. pteronyssinus* measures almost 0.5 mm. So on a surface measuring one square mm there is room for four adult mites. Thus on one square cm there is room for 4.10^2 mites. One square m has room for 4.10^6 mites and one square km can accommodate 4.10^{12} mites. Starting from a single female the population size amounts to 708,516,206 females on day

256 (Fig. 8.1), i.e. 1,417,032,412 male and female mites and eggs: enough to cover a surface of $354\,m^2$. By then, the population has reached a virtually stable age structure and increases according to the formula $N(t) = N(0)e^{rt}$. With $r = 0.073$, it takes only 118 more days to reach a population size of 4.10^{12} and cover one square kilometer. So it takes $(256 + 118=)$ 374 days, i.e. a little more than 1 year, to cover $1\,km^2$ with mites, starting from a single female.

The surface of the earth measures roughly $5.10^8\,km^2$, room for 2.10^{21} mites. It would take 646 days, i.e., less than 2 years, to cover the entire earth, land, and sea, with mites starting from one inseminated female. Then the mite bodies start to pile up. Four months later the layer is 1 m thick. Another 4 months and a thickness of 1 km is exceeded. The rising goes faster and faster—a hallmark feature of exponential growth.

The reason why a catastrophe like this will never happen is, of course, that the three conditions mentioned in the beginning of this section are not fulfilled. These three conditions were optimum temperature, optimum humidity, and food ad libitum.

It is important to realize that the first mentioned factors, temperature and humidity, are density-independent factors. Even a very strong down-regulation of these factors will not prevent a population explosion. With less favorable physical conditions the value of r is smaller, which means that the population increase proceeds more slowly, but it is still an exponential increase. A population explosion is postponed, but it will eventually happen, just as dramatically (see Fig. 8.2). Only if temperature and/or humidity are changed strongly enough to reduce the Malthusian parameter r to zero, the population increase comes to a halt.

Things are entirely different with the third factor: food availability. The essential difference with temperature and air humidity is that the availability of food depends among other things on the population density of the mites. In contrast with temperature and humidity, food availability is a density-dependent factor. The distinction between density-dependent and density-independent factors cannot be overemphasized. When there is plenty of food, mite numbers can increase rapidly. But with more mites the food stocks deplete at a faster rate, leading to a slowing down of the population increase, due to food scarcity. So there is a feedback loop here: more food leads to more mites, but more mites lead to less food. I will address that issue in Section 8.3. But first I will compare the calculated age structures with the available information on the composition of real mite populations.

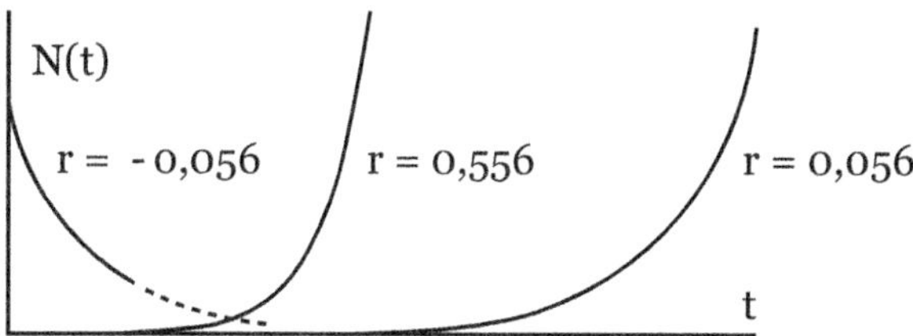

Fig. 8.2 $N(t) = N(0)e^{rt}$. $N(t)$ is the number of mites at time t. $N(0)$ is the initial number of mites, i.e. at time $t = 0$. The parameter r is the "Malthusian parameter."

8.2 A speculative appraisal of empirical findings

(Mathematical details concerning this section can be found in Appendix F.)

The life and fertility table considered in the previous section concerns dust mites living in a land of plenty—a dust mite paradise, created by Arlian and his coworkers. No predators, no dangers imposed by vacuum cleaners, no shortage of food, no drought spells, and an agreeable temperature all the time. Life and fertility tables record the daily survival and egg production of a *cohort* of cultured females. It is practically impossible to obtain such information about a cohort of *D. pteronyssinus* females living in a carpet. But concerning the existing age structure of *populations* living in the house dust habitat, some interesting information is available from two studies: one by Suto et al. (1991) and one by Colloff (1992). Suto et al. determined the numbers of adults, nymphs, and larvae of *D. farinae* and *D. pteronyssinus* in 10 homes in the city of Nagoya (Japan) at three-month intervals during 2 years. The data concerning *D. farinae* were considered before (Section 4.6) in connection with the propensity of this species to remain dormant for a long time. The population dynamics of *D. pteronyssinus* must be very different. Table 8.2. presents information concerning its population structure in Nagoya homes. Nonparametric statistics indicate that the differences in population structure between the seasons are not fortuitous (Kruskal-Wallis test: $.005 < P < .01$). In both years the % adults was lowest in March and highest in June. Perhaps this pattern reoccurs every year. Anyway, at all times the adults greatly outnumbered the immatures. This is in sharp contrast with the calculated age structure described in the previous section. In that rapidly expanding population the adults were a minority and the younger stages dominated: eggs were most numerous, followed by larvae, then protonymphs, then tritonymphs, and then the adults.

Allow me to speculate about an explanation for this contrast in age structure. I guess that the difference in temperature and humidity in the houses will not strongly modify the stable age structure. Under less favorable physical conditions the developments will proceed more slowly. This will retard the development of a stable age distribution but the ultimate age structure is not necessarily influenced.

Table 8.2 Percent adults in *D. pteronyssinus* populations living in carpets and furniture in Nagoya (Japan). Data were read from histograms showing mean numbers of females, males, nymphs, and larvae per 0.1 g of fine dust. Eggs were not recorded.

	December	September	June	March
1st year	83.0	88.4	92.0	88.0
	87.2	82.9	87.9	69.1
	78.2	88.0	81.5	70.9
	86.3	89.0	100.0	79.5
	80.2	89.2	84.7	78.2
	87.3	85.0	100.0	100.0
	84.9	88.0	92.1	
Average	**83.86**	**87.22**	**91.17**	**80.97**
St. dev.	**3.57**	**2.35**	**7.11**	**11.51**
2nd year	85.4	86.6	100.0	80.1
	82.6	86.8	85.5	78.4
	78.3	77.3	89.3	82.9
	77.1	83.5	86.4	77.5
	100.0	82.8	86.5	81.2
	87.2	78.7	78.7	80.8
		84.7	100.0	73.7
Average	**85.10**	**82.92**	**89.49**	**79.22**
St. dev.	**8.28**	**3.69**	**7.87**	**3.02**

Published by Suto, C., Sakaki, I., Ito, H., 1991. Comparative studies on the population dynamics of house dust mites, *Dermatophagoides farinae* and *D. pteronyssinus* in homes in Nagoya, Jpn. J. Sanit. Zool. 42: 129–140.

Vacuum cleaners regularly remove some live dust mites. That has been shown over and over again in the thousands of vacuumed dust samples that have been analyzed acarologically. It is possible to determine the proportion of the mites that is removed and the proportion that remains in a carpet after vacuuming (see Appendix A), but unfortunately nobody seems to have done that. In order to study the effect of increased mortality I made some hypothetical changes in the life table. Table 8.3A presents some fabricated life tables together with the population structures which would develop if it were real. Only the first one (A) is not fabricated but based on the observed data published by Arlian et al. for mites living in near optimum circumstances. These privileged mites gradually die through old age. We do not understand the precise physiological causes of this mortality. I will refer to it as "intrinsic mortality" to distinguish it from the mortality through external causes such as vacuuming and predation. It is important to notice that the "intrinsic mortality" is supposed to operate also in the fabricated life tables B through G, so that also these mites never live longer than 12 periods. When the mortality due to external causes is so high that, in each period, only 55% survives (life table B: $q = 0.55$), then the value of r is very low (row r in Table 8.3A). However, in the calculated age structure (shown at the bottom of the table) the adults still constitute a small minority. So we still don't have an explanation for the preponderance of adults in the Japanese houses. A population like this (with $r = 0.0005$) is barely surviving, i.e., the size of it remains more or less the same. When survival is only a little less (life table C), the value of r turns negative and the population will decline and ultimately disappear.

Fertility seems to have a stronger influence on the ultimate age structure. When the egg production is only 5% of the potential production recorded for near optimum circumstances (life table D: m-fraction $= 0.05$), the adults constitute as much as 41.9% in the calculated population structure—a much greater proportion of adults but still far below the 80% or 90% that is seen in the Japanese furniture. Again, when egg production is only a little less, the value of r turns negative (life table E). In life table F mortality caused by external factors is combined with a reduced fertility, which is more realistic. In the calculated age structure the proportion of adults is 16.6% (25.5% when the eggs are disregarded). So a life and fertility table that leads to a preponderance of adults is still not found.

Things are different when we abandon the assumption that the intrinsic mortality remains the same in all circumstances. In Table 8.3B some hypothetical life tables are shown in which, during a number of periods, the intrinsic mortality among adults is zero. In life table H, the number of living females remains the same during the periods 3 through 10, that is, intrinsic mortality as well as the mortality due to external causes $(1 - q)$ is zero. The total egg production is the same as under near optimum circumstances. In the associated age structure, adults constitute 12% of the population. If mortality remains zero up to period 17, so that females live much longer (life table I), the proportion of adults goes up to 18.6%. Still not very high. But if also the fertility, i.e. the total egg production, is reduced to only 5% (life table J, m-fraction=0.05) the adults become really dominant in the ensuing population structure. So it seems that the dominance of adults in *D. pteronyssinus* populations living in the Japanese houses could be

Table 8.3A Fabricated life tables (B through G) and the associated calculated age structures based on the empirical life table data (A) recorded for *D. pteronyssinus* in (near) optimum conditions. The meaning of "q" and "*m*-fraction" is explained in Table 8.3B (bottom).

		Empirical			Fabricated life tables			
		A	B	C	D	E	F	G
	q	1	0.55	0.5	1	1	0.7	0.7
	m-fraction	1	1	1	0.05	0.01	0.3	0.2
	Pivotal age (periods)	♀♀	♀♀	♀♀	♀♀	♀♀	♀♀	♀♀
Cohort life	0.5	65	65	65	65	65	65	65
table	1.5	63	35	32	63	63	44	44
	2.5	61	18	15	61	61	30	30
	3.5	59	10	7	59	59	20	20
	4.5	57	5	4	57	57	14	14
	5.5	56	3	2	56	56	9	9
	6.5	54	1	1	54	54	6	6
	7.5	43	1	0	43	43	4	4
	8.5	31	0	0	31	31	2	2
	9.5	14	0	0	14	14	1	1
	10.5	4	0	0	4	4	0	0
	11.5	2	0	0	2	2	0	0
	12.5	0	0	0	0	0	0	0
	r (periods)	0.5880	0.0005	−0.0918	0.0645	−0.1914	0.0187	−0.0524
		%	%		%		%	
Popul. age	Eggs	46.6	47.1		16.7		34.7	
structure	Larvae	25.1	25.1		15.2		23.1	
	Proton.	13.5	13.3		13.8		15.4	
	Triton.	7.2	7.1		12.5		10.2	
	Adult	7.7	7.5		41.9		16.6	

Table 8.3B As Table 8.3A, but the intrinsic mortality is relaxed.

Fabricated life tables

		H		I		J		K	
	q	1.0		1.0		1.0		0.9	
	m-*fraction*	1.00		1.00		0.05		0.05	
	Pivotal age	♀♀	♀**eggs**	♀♀	♀**eggs**	♀♀	♀**eggs**	♀♀	♀**eggs**
Cohort		65.0		65.0		65.0		65.0	
life	0.5	63.0	0.0	63.0	0.0	63.0	0.0	56.7	0.0
table	1.5	61.0	0.0	61.0	0.0	61.0	0.0	49.4	0.0
	2.5	59.0	0.0	59.0	0.0	59.0	0.0	43.0	0.0
	3.5	57.0	0.0	57.0	0.0	57.0	0.0	37.4	0.0
	4.5	57.0	257.1	57.0	133.0	57.0	6.7	33.7	12.3
	5.5	57.0	257.1	57.0	133.0	57.0	6.7	30.3	11.1
	6.5	57.0	257.1	57.0	133.0	57.0	6.7	27.3	10.0
	7.5	57.0	257.1	57.0	133.0	57.0	6.7	24.5	9.0
	8.5	57.0	257.1	57.0	133.0	57.0	6.7	22.1	8.1
	9.5	57.0	257.1	57.0	133.0	57.0	6.7	19.9	7.3
	10.5	57.0	257.1	57.0	133.0	57.0	6.7	17.9	6.5
	11.5	0.0	128.6	57.0	133.0	57.0	6.7	16.1	5.9
	12.5	0.0	0.0	57.0	133.0	57.0	6.7	14.5	5.3
	13.5	0.0	0.0	57.0	133.0	57.0	6.7	13.0	4.8
	14.5	0.0	0.0	57.0	133.0	57.0	6.7	11.7	4.3
	15.5	0.0	0.0	57.0	133.0	57.0	6.7	10.6	3.9
	16.5	0.0	0.0	57.0	133.0	57.0	6.7	9.5	3.5
	17.5	0.0	0.0	57.0	133.0	57.0	6.7	8.6	3.1
	18.5	0.0	0.0	0.0	66.5	0.0	3.3	0.0	1.5
	19.5	0.0	0.0	0.0	0.0	0.0	0.0	0.0	0.0
	Total eggs		1928.5		1928.5		96.4		96.4
	r (periods)	**0.5057**		**0.4035**		**0.0360**		**0.0430**	
		%		%		%		%	
Popul. age	Eggs	41.4		34.9		8.0		16.0	
structure	Larvae	24.2		22.6		7.5		13.3	
	Proton.	14.1		14.6		7.0		11.1	
	Triton.	8.2		9.4		6.5		9.3	
	Adults	12.0		18.6		71.0		50.3	

q is the fraction of the mites escaping "external" causes of mortality like vacuum cleaning and predation.
m-fraction (maternity fraction) is the fraction of the potential (maximum) production of female eggs per female.
r is the intrinsic rate of increase (Malthusian parameter).
Note that time is measured in "periods." Only in (near) optimum circumstances (23°C, 75%RH) a period lasts 8 days so that for life table A, r can be divided by 8 to obtain its value with time in days.

explained by a greatly increased longevity combined with a greatly reduced egg production.

It has been shown (Arlian et al., 1990; Matsumoto et al., 1986) that unmated females live longer than mated ones. The inability of unmated females to lay eggs possibly underlies this increased longevity. Perhaps, if not lack of insemination but scarcity of food prevents females from laying eggs, this also leads to a longer life span. Here we have something that can be examined experimentally. Do female dust mites live longer when put on a lean diet? And what about their egg production? Administration of a defined, restricted food supply is not simple. It is much more complicated than subjecting mites to a suboptimal temperature or humidity. But it must be possible. A nice challenge for an experimentalist.

It should also be remembered that Suto et al. (1991) did not record the numbers of eggs. When eggs are ignored the percentage of adults is greater. On the other hand, it should be remembered that the considerations mentioned before concern only females not males. And the longevity of males can hardly be influenced by egg production. If females live much longer than males, this must lead to a female-biased sex ratio. And that is not what we see. The sex ratio recorded by Suto et al. is very close to 1:1. The fraction of females in 54 samples is on average 0.50 (range: 0.3–0.70). So there are some more questions to be answered.

Another set of interesting data was published by Colloff (1992). It concerns mite populations, mainly *D. pteronyssinus* and *E. maynei*, living in eight mattresses in Glasgow (UK). The age structure of *D. pteronyssinus* is quite different from the ones recorded by Suto et al. In these mattresses the immature mites dominate, not the adults. No X^2-tests or confidence intervals are presented, nor the total numbers of mites, needed to calculate these statistics, but the overall picture is that the age structure changes over time, indicating that the populations are evolving. Generally the immature mites dominate. The dust samples were taken on four occasions, 1 month apart, between January and May 1985. A shift toward a dominance of eggs during the latter 2 months concurred with an increasing population density. Also interesting, the age structure of *E. maynei* was entirely different from that of *D. pteronyssinus*. In *E. maynei* the older stages dominated instead of the younger ones.

8.3 A density-dependent factor, namely: Food

Availability of more food will lead to more mites. With more mites the food stock is eaten away faster. Eventually mites are so numerous that the

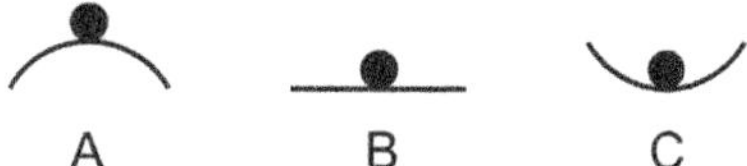

Fig. 8.3 Artistic presentation of a ball on a surface in an unstable situation, in an indifferent position, and in stable equilibrium.

food is eaten away faster than it can be replenished leading to food scarcity and, in its wake, a slower reproduction of the mites. Eventually food supply and population density will balance one another (Fig. 8.3). I used the computer program Stella, a commercially available graphical simulation program, to keep track of the changes in a modeled hypothetical house dust ecosystem, inhabited by one species of house dust mite. This mite has the properties of *D. pteronyssinus* as far as they are known. The information that is needed for the model but not available had to be guessed at. It is nevertheless interesting to see how the model behaves. The diagram shown in Fig. 8.4. speaks largely for itself. Formulas are given in Appendix G.

The model concerns a population of mites in $40\,\text{m}^2$ of fixed, wall to wall carpet of a house that is inhabited by four people. According to Goldschmidt and Kligman (1963, see section 9), each one of these four people will shed 0.5–1 g of crude soft horn each day, roughly 0.75 g on average. When 0.75 g of edible skin debris is shed by each of the four people, each day, this adds up to $0.75 \times 4 \times 7 \times 10^6 = 21 \times 10^6\,\mu\text{g}$ of food per week. If, for instance, 0.2% of this ends up in the $40\,\text{m}^2$ of carpet, then: $0.002 \times 21 \times 10^6/ 40\,\mu\text{g} = 1050\,\mu\text{g/m}^2$ per week is added to the food stock. The rest is lost outside the house or in the mattresses, or it goes down the drain of the shower and the washing machine or it is removed by vacuuming before it penetrates deep into the carpet within reach of the mites and out of reach of the vacuum cleaner. If, instead of 0.2%, as much as 15% ends up in the $40\,\text{m}^2$ of carpet, then 78,750 $\mu\text{g/m}^2$ per week is added to the food stock. Let's run the model for 1000 and 80,000 $\mu\text{g/m}^2$ per week (NET DEPOSITION, Fig. 8.4, and Table 8.4A, column 1). It is good to remember that these values are not only very uncertain, but certainly also very variable among different dwellings. And we are assuming that the food particles are spread out evenly over the surface of the carpet, which is unrealistic. But what else can we do?

The next question we have to answer is: How much food is consumed per mite per week?

Arlian (1977) determined feeding rates of adult female dust mites living in optimum physical and nutritional circumstances. The mites were fed on yeast. The mass of the yeast eaten per day by a single female was 0.60 µg at 25°C and 75%RH and as much as 2.95 µg at 25°C and 85% RH. Let's assume that this is similar to the mass of the skin scales eaten per day per female. Per week this would be 4.20 and 20.65 µg/wk per female, respectively.

A subadult female does not lay eggs but she will have to eat in order to grow and to generate energy for moving and for the extraction of water vapor from the air. In a rapidly growing population, a large number of the individuals are nonfeeding eggs. Thus, on average, much less than 4.20 µg of skin scales may be eaten each week by an average mite living in the land of plenty and in ideal physical circumstances. Let's run the model for 2 and for 15 µg of food per week and per mite (EAT).

Usually circumstances are not ideal for dust mites, certainly not for long. The model must be instructed as to how reproduction and the consumption of food are influenced by a more restricted availability of food and less favorable physical circumstances. If there is no food at all, there can be no reproduction: CONSUMPTION FACTOR $= 0$. We can only guess about the minimum amount of food (in $\mu g/m^2$) that must be present in the carpet in order to allow some reproduction. On the other hand, when the amount of food in the carpet keeps going up, the situation will arise at some point, that adding more food has no effect on the reproduction rate anymore. I will call this the "saturation level" ($\mu g/m^2$). In order to get some faint idea about the position of this saturation level on the scale of food quantity per square meter, I made the following observation: I weighed a little bit of the food that I use for my mite cultures (ground dried yeast and *Daphnia*) and spread it over a sheet of paper. Then I shook it off to another sheet. Microscopic examination learned that the density of food particles remaining on the paper is so high that a dust mite would need a minimum of time and energy to find a food item. I needed 0.895 g of food to contaminate the total surface of 51 sheets of paper (A4; $0.0624\,m^2$). Thus $28 \times 10^4\ \mu g/m^2$ is ample food. Undoubtedly half of this, even much less than half of this, would be enough to allow a maximum reproduction rate. On the other hand it must be more difficult to find the food particles between the piles of a carpet than on a flat sheet of paper.

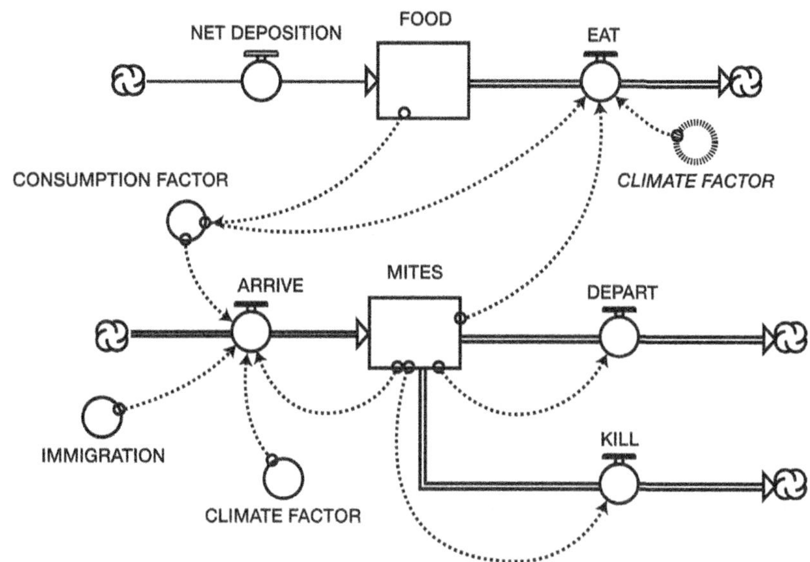

Fig. 8.4 Diagram of the model.

Table 8.4A Values of the parameters of the model.

Set nr.	Net deposition (µg/m²/week)	EAT (µg/week/ mite)	Cons. factor minimum (µg/m²)	Cons. factor Sat. level (µg/m²)	Climate factor	Depart (fraction of mites)
	1	2	3	4	5	6
1	1000	2	0	10,000	0.5	0.01
2	1000	2	0	10,000	0.5	0.06
3	1000	2	0	10,000	0.2	0.01
4	1000	2	0	10,000	0.2	0.06
5	1000	2	0	70,000	0.5	0.01
6	1000	2	0	70,000	0.5	0.06
7	1000	2	0	70,000	0.2	0.01
8	1000	2	0	70,000	0.2	0.06
9	1000	2	5000	10,000	0.5	0.01
10	1000	2	5000	10,000	0.5	0.06
11	1000	2	5000	10,000	0.2	0.01
12	1000	2	5000	10,000	0.2	0.06
13	1000	2	5000	70,000	0.5	0.01
14	1000	2	5000	70,000	0.5	0.06
15	1000	2	5000	70,000	0.2	0.01
16	1000	2	5000	70,000	0.2	0.06
17	1000	15	0	10,000	0.5	0.01
18	1000	15	0	10,000	0.5	0.06
19	1000	15	0	10,000	0.2	0.01

Continued

Table 8.4A Values of the parameters of the model—cont'd

Set nr.	Net deposition ($\mu g/m^2$/week)	EAT (μg/week/mite)	Cons. factor minimum ($\mu g/m^2$)	Cons. factor Sat. level ($\mu g/m^2$)	Climate factor	Depart (fraction of mites)
	1	*2*	*3*	*4*	*5*	*6*
20	1000	15	0	10,000	0.2	0.06
21	1000	15	0	70,000	0.5	0.01
22	1000	15	0	70,000	0.5	0.06
23	1000	15	0	70,000	0.2	0.01
24	1000	15	0	70,000	0.2	0.06
25	1000	15	5000	10,000	0.5	0.01
26	1000	15	5000	10,000	0.5	0.06
27	1000	15	5000	10,000	0.2	0.01
28	1000	15	5000	10,000	0.2	0.06
29	1000	15	5000	70,000	0.5	0.01
30	1000	15	5000	70,000	0.5	0.06
31	1000	15	5000	70,000	0.2	0.01
32	1000	15	5000	70,000	0.2	0.06
33	80,000	2	0	10,000	0.5	0.01
34	80,000	2	0	10,000	0.5	0.06
35	80,000	2	0	10,000	0.2	0.01
36	80,000	2	0	10,000	0.2	0.06
37	80,000	2	0	70,000	0.5	0.01
38	80,000	2	0	70,000	0.5	0.06
39	80,000	2	0	70,000	0.2	0.01
40	80,000	2	0	70,000	0.2	0.06
41	80,000	2	5000	10,000	0.5	0.01
42	80,000	2	5000	10,000	0.5	0.06
43	80,000	2	5000	10,000	0.2	0.01
44	80,000	2	5000	10,000	0.2	0.06
45	80,000	2	5000	70,000	0.5	0.01
46	80,000	2	5000	70,000	0.5	0.06
47	80,000	2	5000	70,000	0.2	0.01
48	80,000	2	5000	70,000	0.2	0.06
49	80,000	15	0	10,000	0.5	0.01
50	80,000	15	0	10,000	0.5	0.06
51	80,000	15	0	10,000	0.2	0.01
52	80,000	15	0	10,000	0.2	0.06
53	80,000	15	0	70,000	0.5	0.01
54	80,000	15	0	70,000	0.5	0.06
55	80,000	15	0	70,000	0.2	0.01
56	80,000	15	0	70,000	0.2	0.06
57	80,000	15	5000	10,000	0.5	0.01
58	80,000	15	5000	10,000	0.5	0.06
59	80,000	15	5000	10,000	0.2	0.01
60	80,000	15	5000	10,000	0.2	0.06
61	80,000	15	5000	70,000	0.5	0.01
62	80,000	15	5000	70,000	0.5	0.06
63	80,000	15	5000	70,000	0.2	0.01
64	80,000	15	5000	70,000	0.2	0.06

Table 8.4B Results of running the model, starting with zero food and zero mites.

Set nr.	Type	Food peak occurrence (weeks)	Food peak maximum ($\mu g/m^2$)	Mite peak occurrence (weeks)	Mite peak maximum (mites/m^2)	Food after 6 yrs ($\mu g/m^2$)	Mites after 6 yrs (mites/m^2)	Food dynamics (see Section 9.2)	Class (effect of KILL)
	1	2	3	4	5	6	7	8	9
1	1	24	20,442	no	–	401	25,077	a	A
2	2	29	24,056	41	7069	2304	4361	a	C
3	1	59	48,999	no	–	987	25,460	a	A
4	2	108	86,421	150	15,860	5761	4361	a	B
5	1	51	44,974	no	–	2778	25,345	b	A
6	2	62	54,541	86	9394	16,140	4362	b	C
7	1	83	72,928	no	–	6765	26,011	b	A
8	2	139	117,078	183	14,054	46,678	3887	a	B
9	1	26	21,693	no	–	5201	25,000	a	A
10	2	30	25,342	41	6625	6152	4361	a	C
11	1	60	49,751	no	–	5496	25,351	a	A
12	2	109	87,201	151	15,778	7880	4361	a	B
13	1	53	47,234	no	–	7588	25,263	b	A
14	2	63	56,645	87	9149	19,983	4362	b	C
15	1	85	74,326	no	–	11,312	25,897	b	A
16	2	140	118,431	184	14,061	47,936	4023	a	B
17	1	16	12,937	no	–	392	3423	b	A
18	2	19	14,746	32	678	2249	596	a–b	C
19	1	38	28,872	no	–	974	3440	a	A
20	2	64	45,489	96	1386	5621	596	a	C
21	1	41	33,253	no	–	2723	3448	b	A
22	2	51	41,203	78	990	15,742	596	b	C

Continued

Table 8.4B Results of running the model, starting with zero food and zero mites—cont'd

Set nr.	Type	Food peak occurrence (weeks)	Food peak maximum (μg/m^2)	Mite peak occurrence (weeks)	Mite peak maximum (mites/m^2)	Food after 6 yrs (μg/m^2)	Mites after 6 yrs (mites/m^2)	Food dynamics (see Section 9.2)	Class (effect of KILL)
	1	2	3	4	5	6	7	8	9
23	1	66	53,861	no	–	6712	3500	b	A
24	2	96	76,965	135	1158	39,130	635	b	A
25	1	18	14,181	no	–	5196	3413	a	A
26	1–2	20	16,002	33	605	6124	596	a	C
27	1	39	29,648	no	–	5489	3426	a	A
28	2	65	46,264	96	1375	7811	596	a	A–B
29	1	43	35,734	no	–	7536	3438	b	A
30	2	53	43,727	79	965	19,616	596	b	C
31	1	68	55,554	no	–	11,258	3486	b	A
32	2	97	78,273	136	1155	41,090	629	b	A
33	1	40	2,844,586	no	–	401	2,006,460	a	A
34	2	48	3,465,213	61	967,439	2313	347,586	a	C
35	2	104	7,509,182	132	2,469,766	935	2,149,016	a	A
36	2	205	14,622,148	261	2,260,641	4597	436,900	g	B
37	1	40	2,847,492	no	–	2805	2,006,492	a	A
38	2	48	3,467,705	61	967,760	16,192	347,586	a	C
39	2	104	7,510,250	132	2,468,852	6548	2,148,794	a	A
40	2	205	14,623,266	261	2,261,219	32,175	435,104	g	B
41	1	40	2,844,626	no	–	5492	2,006,221	a	A
42	2	48	3,465,247	61	966,293	6157	347,586	a	C
43	2	104	7,509,195	132	2,468,581	5468	2,148,826	a	A
44	2	205	14,622,162	261	2,259,506	7299	436,996	g	B

Table 8.4B Results of running the model, starting with zero food and zero mites—cont'd

Set nr.	Type	Food peak occurrence (weeks)	Food peak maximum (μg/m^2)	Mite peak occurrence (weeks)	Mite peak maximum (mites/m^2)	Food after 6 yrs (μg/m^2)	Mites after 6 yrs (mites/m^2)	Food dynamics (see Section 9.2)	Class (effect of KILL)
	1	2	3	4	5	6	7	8	9
45	1	40	2,847,667	no	–	7605	2,006,398	a	A
46	2	48	3,467,852	61	966,793	20,035	347,586	a	C
47	2	104	7,510,313	132	2,467,683	11,081	2,148,607	a	A
48	2	205	14,623,331	261	2,260,175	34,881	435,207	g	B
49	1	32	2,201,196	no	–	402	266,935	a	A
50	2	38	2,662,732	50	105,509	2312	46,359	a	C
51	1–2	83	5,801,811	108	270,831	967	277,204	a	A
52	2	160	10,990,770	210	245,769	5726	46,802	g	B
53	1	32	2,204,138	no	–	2811	266,940	a	A
54	2	38	2,665,249	50	105,472	16,187	46,359	a	C
55	1–2	83	5,802,919	108	270,643	6768	277,183	a	A
56	2	160	10,991,915	210	245,603	40,110	46,760	g	B
57	1	32	2,201,237	no	–	5413	266,907	a	A
58	2	38	2,662,766	50	105,359	6156	46,359	a	C
59	1–2	83	5,801,826	108	270,675	5483	277,183	a	A
60	2	160	10,990,785	210	245,767	7863	46,805	g	B
61	1	32	2,204,314	no	–	7611	266,928	a	A
62	2	38	2,665,398	50	105,343	20,031	46,359	a	C
63	1–2	83	5,802,984	108	270,496	11,285	277,163	a	A
64	2	160	10,991,981	210	245,595	42,243	46,763	g	B

Let's run the model for two different assumed saturation levels: 1×10^4 and 7×10^4 μg/m^2, each one combined with two different "minimum values," namely: 5×10^3 μg/m^2 and 0 μg/m^2. The CONSUMPTION FACTOR in this model depends entirely on the availability of food. It influences both the rate of food intake (EAT) and the rate of reproduction (ARRIVE). I assumed that these influences are equivalent, i.e., when the consumption of food is decreased by, for instance, a factor 0.2, then the reproduction rate is also decreased by a factor 0.2.

The availability of food is one of the two factors that influence reproduction. The other one is the climate, i.e., temperature and RH. The CLIMATE FACTOR also ranges from 0 (no reproduction, no food intake) to 1 (optimum). Again it is assumed that its influence on reproduction is equivalent to its influence on food intake. Let's assume for now that there are no seasons; the climate provides constant atmospheric conditions. But these conditions are not optimal. For one series of runs the CLIMATE FACTOR is set 0.5 and for another series it is set 0.2.

The fraction of the mites that are removed by the vacuum cleaner each week must depend very much on the intensity of vacuuming. Let's run the model for two assumed fractions: 1% and 6% (DEPART).

Finally, for a population to get started there must be some immigration of mites from elsewhere. Let's assume that each week one mite per square meter is imported by passive dispersal (IMMIGRATION).

Though pervaded with uncertainties, the model demonstrates possible scenarios. Two values for each of six parameters make 64 possible sets of conditions. I ran the model for each one of them. Tables 8.4A and 8.4B give an overview of the results. Three examples are shown graphically in Fig. 8.5. Here the development of the food stock and mite density is depicted from installation of the new carpet until 6 years later. The food density curve invariably starts with a spike at the beginning, caused by the initial absence of food consuming mites. The curves of the mite density developments are less uniform. Two types can be distinguished. In TYPE 2 there is an initial, short lived, peak in the number of mites (Fig. 8.5, top). This happens when other factors besides food availability suppress the increase of mite numbers, for instance low temperatures, low air humidity, or high mortality due to intense vacuum cleaning. When at last the number of reproducing females becomes significant, food has accumulated so much that the increase rate rises to a high value. But not for long because, with so many mites, the food stock is quickly exhausted and after a short lived peak, mite numbers fall again. When, in contrast, physical circumstances

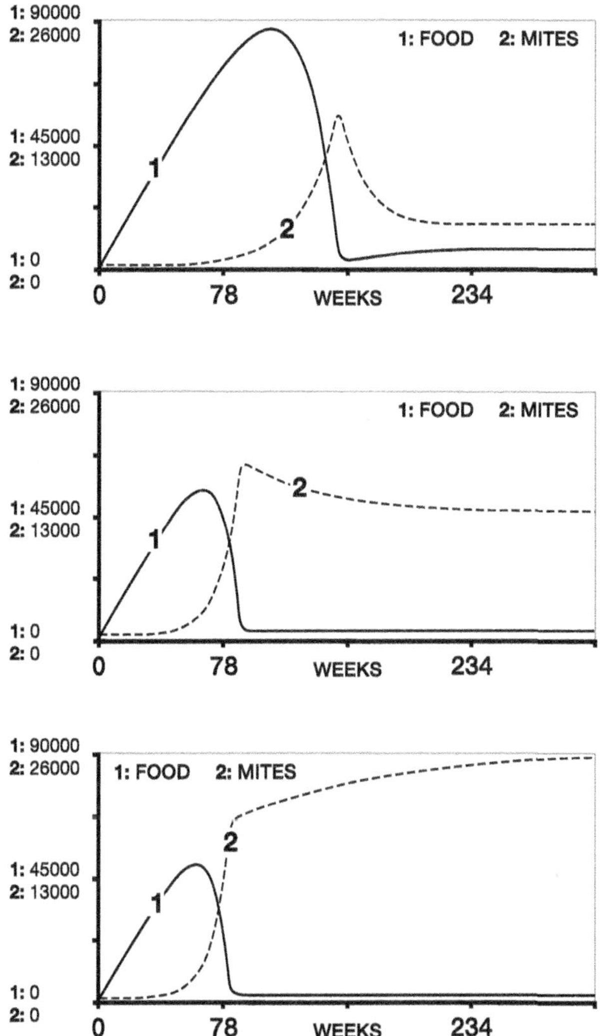

Fig. 8.5 Development of (1) food stock ($\mu g/m^2$) and (2) mite density (mites/m^2) during a 6-year period. Sets Nr. 3 and 4 are specified in Table 8.4A. *Top figure:* Set Nr. 4. DEPART = 6%/week. Example of Type 2 (Tables 8.4B, column 1), i.e., the food stock declines after reaching a maximum. *Bottom figure:* Set Nr. 3. DEPART = 1%/week. Example of Type 1, i. e., the food stock is continuously rising. *Middle figure:* DEPART is 2%, otherwise as the sets Nr. 3 and 4. (See also Fig. 8.4.)

are favorable and mortality is low, food availability is a limiting factor all along. In such cases, there is no initial peak in the mite numbers. The population density rises continuously, albeit ever more slowly. Set Nr. 3 (Fig. 8.5, bottom) is an example. I designated cases without an initial peak in mite numbers as TYPE 1 (Table 8.4B, column 1).

The runs cover a six-year period. After 6 years mite numbers and food densities approach an equilibrium value and fluctuations are only small. The mite numbers per square meter range from 596 (Nr. 22) to 2,148,607 (Nr. 47).

How about the real world? There are plenty of studies reporting numbers of mites per gram of dust but to the best of my knowledge, none ever tried to determine the total number of mites and eggs per square meter of a floor. I once obtained myself an estimate of the total number of *mobile* mites on the carpeted floor of the living room of a private house (de Boer and Kuller, 1995 and Appendix A). Immobile mites and eggs were not recorded because we used the heat escape method. The estimated number of mobile mites per square meter was 8385. If 10% is added to account for the immobile mites and eggs (see Section 8.2), then the mite density would be 9224 mites/m^2. This is intermediate between the values reached by Nr. 2 and Nr. 11 (Tables 8.4A and 8.4B). If 30% is added to account for the immobile stages, the mite density would be 10,900 mites/m^2, which is also intermediate between Nr. 2 and Nr. 11. Part of the carpet was stored for two months at, for dust mites, very favorable physical conditions. The estimated mite density rose to a much higher value, intermediate between those of Nr. 7 and Nr. 50.

So it seems that at least some of the mite population densities produced by the model are not unrealistic.

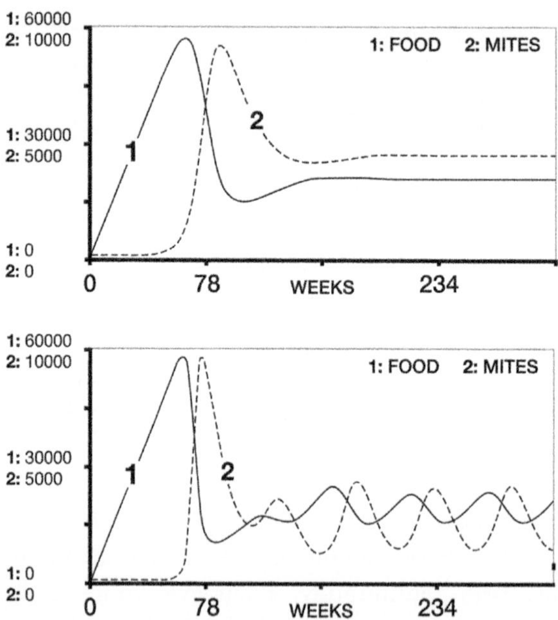

Fig. 8.6 Development of (1) food stock (μg/m^2) and (2) mite density (mites/m^2) during a 6-year period. Set Nr. 14 (specified in Tables 8.4A, column 1). *Top figure:* without seasonal fluctuations (factor SEASON = 1). *Bottom figure:* with seasonal fluctuations (factor SEASON cycles between 0 and 1). (See also Fig. 8.4.)

8.4 Kill them all in winter?

Let us introduce a refinement, namely SEASON. The CLIMATE FAC-TOR is no longer a constant 0.5, but fluctuates in a sinusoidal fashion, reaching 1 in summer and 0 in winter. Estival peaks in mite numbers ensue (Fig. 8.6).

Now suppose the inhabitants of the house hire a team of mite killers every year. These mite killers use agents like liquid nitrogen (see Colloff, 1986), steam (see Colloff et al., 1995), and acaricides (without residue action) and they kill 95% of all the mites. They repeat this action again after exactly 1 year and then 1 year later, etc. For any one of the 64 sets of conditions, the effect of the yearly killing acts is different. But often the differences are only small and only three distinct patterns can be distinguished: CLASS A, B, and C (Fig. 8.7 and Table 8.4B, column 9).

Patterns of CLASS A have the following characteristics: After the initiation of the yearly killings, mite density peaks are (1) lower than before but (2) not decreasing and (3) the food density spikes and slumps (Fig. 8.7A).

In patterns of CLASS B, the mite density peaks after the initiation of the killings are (1) lower than before and (2) decreasing year after year whereas (3) the food density rises continuously (Fig. 8.7B). The food density keeps rising in regions where it is completely out of reach of the vacuum cleaner. Whether such regions really exist we don't know. Perhaps only on the bottom of very thick carpets and not at all in thin carpets. But the mite population will decline anyway.

The killings are repeated at exactly the same time of the year. Wouldn't the timing of these killing actions make a difference, you may ask? I varied the starting time 12 times with a shift of 1 month. When the initiation of the killings is shifted to a different time of the year, the characteristics of both CLASS A and CLASS B remain the same. So in these two classes the season has no substantial influence. That is different for CLASS C. In patterns of CLASS C, the mite killing actions sometimes lead to peaks in mite numbers that, paradoxically, exceed the previous estival peaks in height. The cause of this effect is the increased amount of available food per mite as a consequence of the accumulation of food ensuing from the temporary reduction of mite numbers. The occurrence of peaks exceeding the previous estival peaks depends on the timing of the killing actions. So, in contrast with CLASS A and B, substantial differences in the effect of the mite killing actions may result from a different timing. An example of a CLASS C pattern is shown in Fig. 8.7C and D.

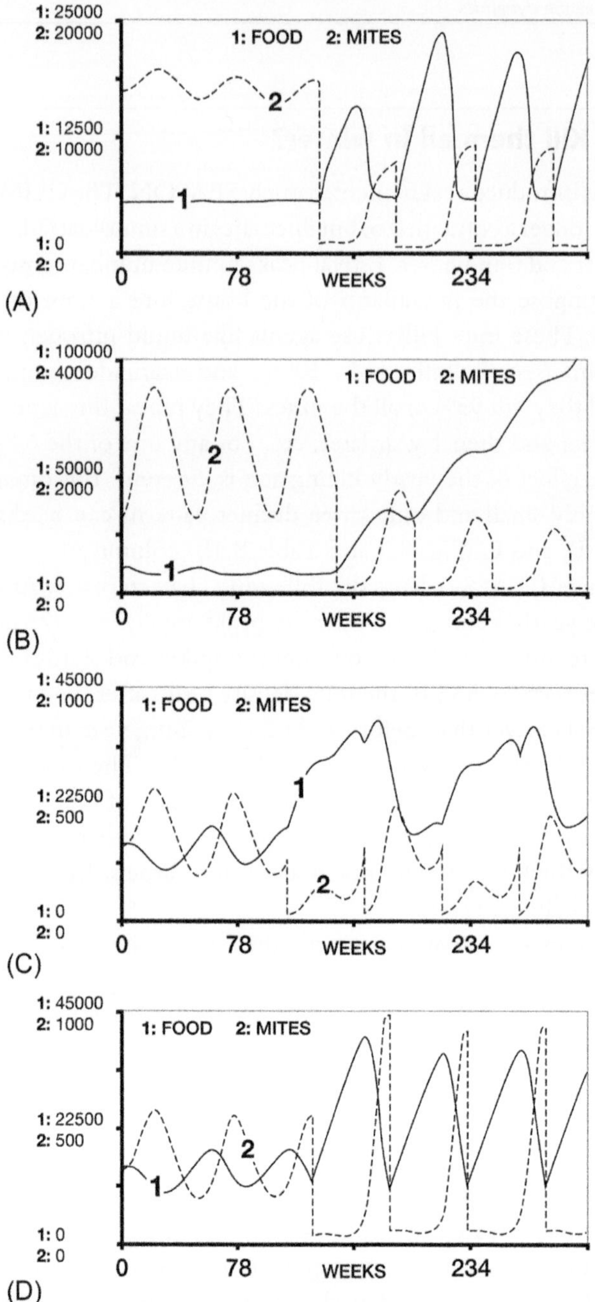

Fig. 8.7 Development of (1) food stock ($\mu g/m^2$) and (2) mite density (mites/m^2) during a 6-year period. Once per year 95% of mites are killed. Specifications of Sets Nr. 11, 12 and 22 in Table 8.4A. (A) Set Nr. 11. Example of Class A (Tables 8.4B, column 9). KILL actions in the weeks 132 + 52n. (B) Set Nr. 12. Example of Class B. KILL actions in the weeks 144 + 52n. (C) and (D) Set Nr. 22. Example of Class D. The only difference between (C) and (D) is in the timing of the KILL actions: a shift of 16 weeks. Description of Class A, B, C, and D in the text.

The pattern of CLASS B is the dreamed scenario. Not for the mites, of course, but for the allergy patients. Mites progressively disappear when, once per year, the population is drastically reduced. The population reduction may be executed at any time of the year, but it is easier to do during the winter. However, a yearly population reduction alone is not enough. The CLASS B scenario occurs only when indoor climate is relatively unfavorable throughout the year (CLIMATE FACTOR $= 0.2$ instead of 0.5 in combination with a relatively high mortality rate: DEPART $= 0.06$ instead of 0.02).

Perhaps in some households, the inhabitants unintentionally create an indoor climate that prevents the development of a thriving mite population. Perhaps that explains the absence of mites in some houses. Unfortunately, adequate observations of indoor climate in such houses are lacking. Theoretically, it is certainly possible to intentionally create such conditions. Whether that is practically feasible and acceptable for the inhabitants will strongly depend on the outdoor climate in the geographic region concerned.

References

Arlian, L.G., 1977. Humidity as a factor regulating feeding and water balance of the house dust mites *Dermatophagoides farinae* and *D. pteronyssinus*. J. Med. Entomol. 14, 484–488.

Arlian, L.G., Rapp, C.M., Ahmed, S.G., 1990. Development of *Dermatophagoides pteronyssinus*. J. Med. Entomol. 27, 1035–1040.

Birch, L.C., 1948. The intrinsic rate of natural increase of an insect population. J. Anim. Ecol. 17, 15–26.

Colloff, M.D., 1986. Use of liquid nitrogen in the control of house dust mite populations. Clin. Allergy 16, 41–47.

Colloff, M.D., 1992. Age structure and dynamics of house dust mite populations. Exp. Appl. Acarol. 16, 49–74.

Colloff, M.D., Taylor, C., Merrett, T.G., 1995. The use of domestic steam cleaning for the control of house dust mites. Clin. Exp. Allergy 25, 1061–1066.

de Boer, R., Kuller, K., 1995. Winter survival of house dust mites (*Dermatophagoides* spp.) on the ground floor of Dutch houses. Proc. Exp. Appl. Entomol. N.E.V Amsterdam 6, 47–52.

Goldschmidt, H., Kligman, A.M., 1963. Quantitative estimation of keratin production by the epidermis. Arch. Dermatol. 88, 709–712.

Matsumoto, K., Okamoto, M., Wada, Y., 1986. Effect of relative humidity on life cycle of the house dust mites, *Dermatophagoides farinae* and *D. pteronyssinus*. Jpn. J. Sanit. Zool. 37, 79–90.

Suto, C., Sakaki, I., Ito, H., 1991. Comparative studies on the population dynamics of house dust mites, *Dermatophagoides farinae* and *D. pteronyssinus* in homes in Nagoya. Jpn. J. Sanit. Zool. 42, 129–140.

CHAPTER NINE

What can dust mites eat?

We know that dust mites can eat particles of organic substance and thrive on it. Cultures of dust mites can be fed with, for instance, a mixture of dried yeast and *Daphnia*. When the mixture is ground to a powder more particles of the right size are formed. In the early days of the dust mite investigations epoch, many other foods have been tried and compared for their suitability to feed dust mite cultures: potato powder, dried beef scrapings, powdered dried blood plasma, powdered milk, fishmeal, and many others, pure or mixed in various combinations. The objective of these early studies was primarily to work out a protocol for bulk rearing dust mites in order to produce allergens for diagnosis or treatments and also to enable the study of dust mite biology. Biological studies focused primarily on the humidity requirements, not on the nutritional requirements. Yet, food is an important aspect of dust mite ecology. Healthy, well fed mites live and breed in our carpets and beds. What their staple food is, none knows exactly. We can only guess.

Sinha et al. (1970) published photographs of a mounted *D. pteronyssinus*. One had a fungal spore (*Alternaria alternata*) and one had a pollen grain (*Iva xanthifolia*) inside its body, presumably inside the gut. But it is not sure whether the mites can digest pollen or fungal spores. Another way to examine the gut content is by squashing a mite on a microscope slide and trying to identify the constituents. van Bronswijk (1973) writes: "The alimentary canal of *D. pteronyssinus* was studied in 147 specimens caught alive on the bedroom floors and 470 specimens caught alive in mattresses. Pollen, spores of microorganisms, fungal mycelia, bacteria and fibers of plant origin (most likely originating in cotton bed sheets) were found frequently during the entire year." In Colloff's book *Dust Mites* (Colloff, 2009) we read: "Squash-preparations of midgut contents of *D. pteronyssinus* taken from freshly sampled house dust contain a variety of components that can be seen under a compound microscope. These include macerated skin scales (indicated by the birefringence of keratin under polarized light), fungal hyphae and spores, yeasts and bacteria, as well as many unidentifiable particles."

Douglas and Hart (1989) made serial paraffin-wax sections of the entire bodies of 26 adult female *D. pteronyssinus* from a culture. In 18 (69%) of the mites they found a total of 74 fungal spores, all of them in the lumen of the

House Dust Mites
https://doi.org/10.1016/B978-0-443-19111-4.00009-5

gut. These findings suggest that the fungi are no permanent symbiotic associates but rather that they were ingested by the mites. However, an extensive bacterial microflora was also present. In every mite examined these bacteria were closely associated with the microvilli of the midgut epithelial cells. Perhaps these bacteria were permanent symbionts. In contrast to the fungi, these bacteria did not grow on nutrient agar.

9.1 Dander

Dander. In my own language (Dutch, as you may have gathered) we don't have a distinct word for it. We use a descriptive designation: "*huidschilfers*," which means "skin scales."

Skin scales are usually considered as the most important food for dust mites. Our skin is continuously worn. The debris that results from it has been called "dander." Goldschmidt and Kligman (1963) made a courageous effort to determine the quantity of dander that is shed from our bodies every day. They acquired the cooperation of 30 inmates of a prison in Philadelphia (USA). These prisoners volunteered to wear stainless steel cylinders, four centimeters in diameter, at various parts of their bodies. These remained there uninterruptedly during periods of 4–6 weeks and in a few cases 2–6 months. Fine steel mesh and gauze on the top of the cylinders prevented the escape of the skin debris. All the sebum-saturated horny material from the confined area was harvested and weighed. Subsequently it was extracted in ether to free it of sebum, a procedure which caused up to 50% weight loss. According to the authors, this brief extraction does not remove lipoids incorporated in the horn itself. After it was washed in warm water and dried, the residue was designated "crude, soft horn." Crude, soft horn contains approximately 70% "insoluble soft keratin." By extrapolation it was concluded that an individual loses 0.5 to 1.0 g of crude, soft horn every day. Microscopic examination of the residue revealed aggregates of cells with intact cell membranes. The size of these aggregates varied between 0.25 mm and several millimeters in diameter (Kligman, 1964), i.e. larger than a dust mite. In a previous section (Section 2.1) I already mentioned that morphological studies indicate that dust mites can swallow particles up to the size of 20 µm. So it seems that freshly shed skin scales need to undergo some changes before they can be eaten by dust mites.

In several studies skin scales were tested for their suitability as a feed for mite cultures in comparison with other products. Unfortunately, details on how these skin scales were obtained and processed are scarcely given. Spieksma (1967) obtained skin scales "from hair collected in barbershops." Besides human skin scales seven other products were tested singly: (1) heart infusion agar, (2) powdered milk, (3) keratin, (4) wheat germ flakes, (5) fishmeal, (6) powdered yeast, and (7) fine dust. The best results were obtained with the skin scales. He also noted that the addition of powdered yeast favored the growth of the mite population significantly.

Hart and Le Merdy (1987) tested 11 foods, both singly and in various combinations: (1) insect meal, (2) wheat germ, (3) dried milk, (4) flour, (5) beard shavings, (6) skin scales, (7) dried liver, (8) fishmeal, (9) dried yeast, (10) vitamin powder, and (11) bran. Rearing on a single food resulted in poor population growth with the exception of fishmeal and insect meal. In contrast to Spieksma's results, skin scales were not so good. Not only *D. pteronyssinus* was examined but also *D. farinae* and *Euroglyphus maynei*. Interestingly, the most favorable single food was different for each of these species.

More recently Molva et al. (2019) also found that an optimal diet for *D. pteronyssinus* is not optimal for *D. farinae*, vice versa. Mixtures of two or more foods often resulted in much better population growth. Also noteworthy was the finding that small particle size seemed to be unfavorable. Hart and Le Merdy suggest that this may be related to the aeration of the mite cultures.

van Bronswijk and Sinha (1973) obtained fresh human dander by combing the scalps of about 100 university students. Part of this dander was defetted in ether for 2 min. Part of this was enriched with yeast powder. Part of this was inoculated with fungus spores in the following way: The mixture of dander and yeast was spread out in a petri dish. A sporulating culture of fungus was stirred vigorously in a tube with 3-mL water. One milliliter of this was added to the petri dish and left for 2–3 weeks at 25°C and 75%RH until the fungi had overgrown the food and the water was evaporated. Then five males and five females of *D. pteronyssinus* were transferred to each dish. The results are shown in Table 9.1. Although the mite numbers in Table 9.1 are the means of as many as 10 replicates, no *P*-values or standard deviations are given. It is strange that the journal allowed this omission. Anyway, it seems that fresh skin scales are a poor food for dust mites. Defetted skin scales are a little better. Addition of other nutrients further improves the quality.

Table 9.1 Number of *D. pteronyssinus* when 10 (5♀+5♂) mites were artificially inoculated in different diets after 4 and 8 weeks at 25°C and 75%RH.

Food	Mean number of mites[a]	
	After 4 weeks	After 8 weeks
Fresh human dander	0	0
Defetted human dander	21	9
Defetted human dander/yeast	38	66
Defetted human dander/yeast +*Aspergillus amstelodami*	93	353
Defetted human dander/yeast +*Aspergillus niger*	25	>5
Defetted human dander/yeast +*Aspergillus tamarii*	37	>5

[a] Mean of 10 replicates.

From van Bronswijk, J.E.M.H., Sinha, R.N., 1973. Role of fungi in survival of Dermatophagoides in house-dust environment. Environ. Entomol. 2, 142–145.

9.2 Food dynamics

When food is not in short supply it means that the amount of food per square meter is above the "saturation level" as defined in Section 8.3. Adding more food would have no influence on development and reproduction. In the models described in Section 8.3. the food density peaks shortly after the installation of a new carpet (Fig. 9.1). When the deposition rate is high (NET DEPOSITION = 80.000 μg/m^2 per week) the initial food peak always rises above the saturation level. In some cases a peak above the saturation level occurs in every subsequent year. These cases are designated g in Table 8.4 (column 8). As you can see in Table 8.4 type g occurs in every set that combines a high deposition rate of food with an unfavorable climate and high mortality. The others are designated a or b. When the deposition rate of food is low (NET DEPOSITION = 1000 μg/m^2 per week) the initial food peak in some cases remains below the saturation level all the time. These cases are designated b. In the cases designated a, the initial food peak rises above the saturation level but afterward the food density remains below it. So according to these models, food for mites is most of the time in short supply.

There is very little observational evidence showing that food supply really is a constraint for the development of mite populations in the stuffings of our homes. In 1989 I performed myself a study that yielded some relevant information. The results were published in a journal (de Boer, 1990). I will present some of the data here, but in a somewhat different way. It was an investigation into the effect of cleaning on the allergen

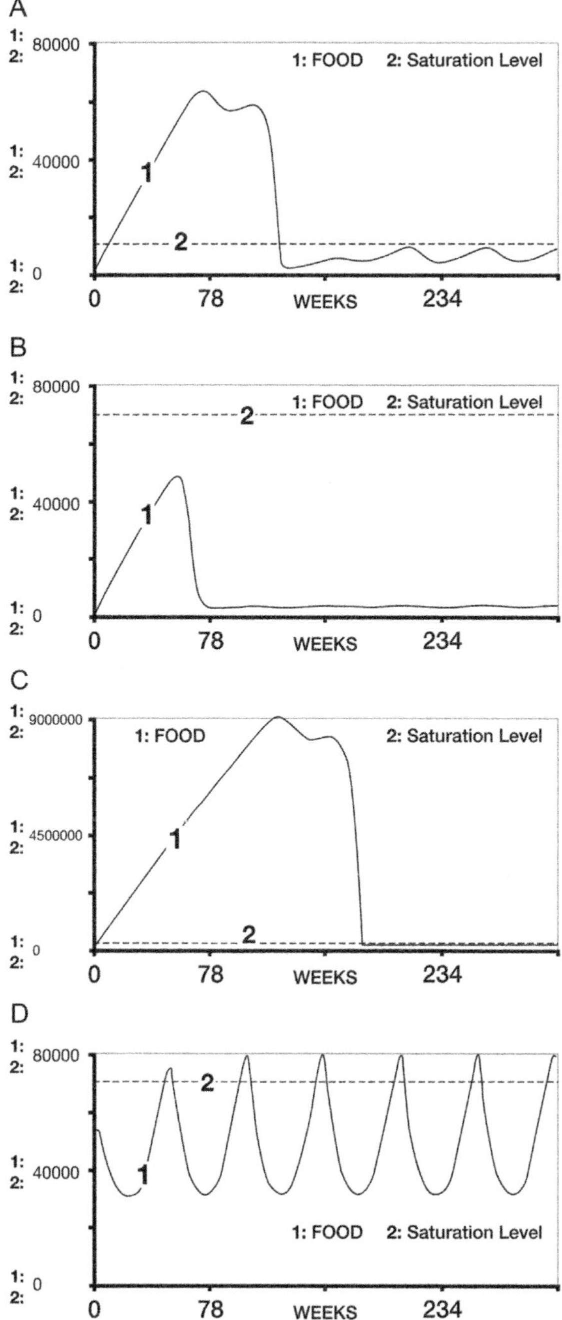

Fig. 9.1 Development of the food stock ($\mu g/m^2$) during a six year period. Sets Nr. 4, 5, and 64 are specified in Table 8.4. (A) Set Nr 4 (example of a, Table 8.4B column 8); (B) Set Nr. 5 (example of b); (C and D) Set Nr. 64 (example of g); (D) is a continuation of (C) for 6 more years. Note the different y-scale. The letters a, b, and g indicate differences in the dynamics of the developing food stock as described in the text.

content of rugs—not only the direct effect of removing the allergens, but also the indirect effect that may ensue from removing mites and food for mites. The six old carpets that were subjected to the investigation had been on the floor of a living room or a bedroom until just before the start of the study. Initially a small vacuumed dust sample was taken to get some information of the resident mite fauna. *D. pteronyssinus* was the dominant species in four of the rugs. In two of the rugs mites belonging to the *D. farinae* group were more numerous. Each rug was cut into four equal parts and from each part 15 fragments, measuring $5 \times 15\,\text{cm}$, were cut out, to serve as untreated controls. The carpet parts, with holes in it, were subjected to either of four cleaning treatments. Vacuum cleaning for an equivalent of $12\,\text{min per m}^2$ was one of the treatments under study. After the treatment, fragments ($5 \times 15\,\text{cm}$) were cut out, adjacent to the holes left by the removal of the control samples. Thus paired comparison of treated and untreated samples is possible. Five sample pairs were used to determine the number of live mites. The heat escape method was applied for this. We already know from numerous other studies that some live mites can be removed by vacuuming. But now we could see that many mites remain in the carpet after it is vacuumed. The numbers are too small to make an assessment of the percentage removed. Probably it was only a small percentage, because the difference in mite numbers of treated and untreated fractions is not significant. From each carpet part five sample pairs were used to test the "habitat suitability." In order to do this, the resident mite population was killed by a cold treatment (48 h at $-25°\text{C}$). Eighty mites were placed on each carpet fragment. The fragments were wrapped in a dust bag and placed in a climatic room at $25°\text{C}$ and 75%RH. After either 3, 4, 5, 6, or 7 weeks, the number of live mites in the treated and the corresponding untreated carpet fragment was compared using the heat escape method. Generally, more mites were recovered from the control samples than from the treated samples. The difference is highly significant (Table 9.2). The most probable explanation for this difference is that food for mites had been removed from the vacuumed samples, indicating that scarcity of food can be a limiting factor for the speed of population increase. The effect of three other cleaning treatments was also tested this way. For each of these cleaning treatments on average fewer mites were recovered from the treated samples than from the untreated controls (de Boer, 1990).

Another 21 (3×7) sample pairs were used to test the effect of a heat treatment (10 min at $110°\text{C}$ in an autoclave). The effect was very strong and highly significant (Table 9.3). By this treatment food is not removed but

Table 9.2 Population development of *D. farinae* in rugs after vacuuming for 12 min/m².

Weeks of incubation	Rug 1 vac.	Control	Rug 1 vac.	Control	Rug 2 vac.	Control
3	41	22	1	2	0	11
4	63	178	78	566	5	8
5	163	138	2	145	2	0
6	109	75	2	108	32	13
7	41	105	7	62	3	1

Weeks of incubation	Rug 3 vac.	Control	Rug 4 vac.	Control	Rug 5 vac.	Control
3	0	3	18	65	3	29
4	2	8	3	14	10	40
5	2	2	24	39	0	3
6	1	1	32	87	12	12
7	0	0	5	38	1	23

Weeks of incubation	Rug 6 vac.	Control
3	205	249
4	141	154
5	117	140
6	87	203
7	167	157

Vac., vacuumed.

Wilcoxon signed-ranks test, one-sided: $P < .001$.

From de Boer, R., 1990. The control of house dust mite allergens in rugs. J. Allergy Clin. Immunol. 86: 808–814.

Table 9.3 Population development of *D. farinae* in rugs after a heat treatment (10 min. at 110°C).

Weeks of incubation	Rug 1 heated	Control	Rug 1 heated	Control	Rug 2 heated	Control
3	1	0	6	47	9	170
4	1	38	0	2	5	96
5	0	0	0	8	0	41
6	0	40	0	10	2	209
7	0	0	0	0	0	209

Wilcoxon signed-ranks test, one-sided: $P \approx .001$.

From: de Boer, R., 1990. The control of house dust mite allergens in rugs. J. Allergy Clin. Immunol. 86: 808–814.

apparently it is rendered unsuitable. That is evidence of the existence of essential but heat unstable food constituents. Possibly a good starting point for research into the food requirements of dust mites.

So there is reason to believe that food for dust mite populations is usually in short supply. However in the cabinets where mites were kept to record their development, reproduction and longevity in order to construct life and fertility tables, food was always available ad libitum. That is a pity. Exposing the mites to suboptimal humidity and suboptimal temperatures is much easier than subjecting them to a suboptimal food supply. That must have been the reason why none ever did that. But it would have been very interesting.

9.3 Fungi and bacteria

Besides dust mites, fungi can be found in house dust. The presence of fungal diaspores does not necessarily mean that the fungi were actually growing and reproducing. But the house dust habitat is a xeric habitat so we would expect only xerophilic fungi to be able to survive. A fungus is regarded as xerophilic if it grows on media containing 30–50% sucrose. The fungi that van de Lustgraaf (1978a) and van de Lustgraaf and Jorde (1977) cultured from mattress dust were indeed xerophilic species belonging to the genera *Aspergillus*, *Eurotium*, *Wallemia*, and *Penicillium*. It is conceivable that the presence of fungi in the house dust ecosystem is a necessary condition for the well-being of house dust mites. Two propositions have been brought forward. Firstly, fungi may provide essential nutrients. Secondly, fungi may have a role in rendering human skin scales more suitable as a food. van Bronswijk and Sinha (1973) used the word "predigestion." Of course the two propositions are not mutually exclusive. Several studies have been done aiming at the elucidation of these questions. For most of these studies the mites were placed in very artificial situations, widely different from the conditions encountered in the house dust ecosystem. Both the quality and the quantity of the food that was offered and the density of the fungi that the mites were exposed to were very unlike the situation in the normal house dust habitat. It is difficult to maintain specified test conditions when living fungi are part of the medium. Replicate ("identical") experiments sometimes yielded significantly different results. Conclusions about the effect of fungi for the mite population growth that could be drawn from the experiments were often ambiguous. Detrimental effects were recorded, especially when fungal growth was excessive. But the same fungus species in a lower density sometimes seemed to have a favorable effect.

Douglas and Hart (1989) isolated and cultured the xerophilic fungus *Aspergillus penicillioides* from spores in the gut of *D. pteronyssinus*. They went on to study the effect of the presence of this fungus in mite cultures

(Hay et al., 1993). To start with, they set up fungus-free cultures of *D. pteronyssinus*. These were obtained from eggs and the hatchlings were propagated on a fungus-free medium. A large sample of the mites were checked for the presence of fungi and deemed fungus free. Then subcultures were set up and propagated for two generations on media with and without fungi. Hay et al. (1993) write in the abstract section of their article: "This fungus reduced survival, development rate, adult length and fecundity of *D. pteronyssinus*. Detrimental effects of *A. penicillioides* were proportional to the fungal density. Despite the antagonistic effects of *A. penicillioides*, a requirement for the fungus was indicated by the poor performance of fungus-free mites in the second generation; sustained culture of *D. pteronyssinus* in the absence of fungi is probably not possible."

Yet another interesting observation was reported in this article. Douglas and Hart acquired the cooperation of seven volunteers who supplied them with their beard shavings. These beard shavings were acetone washed and plated onto agar. The authors write: "With the exception of one volunteer, whose beard shavings were consistently fungus-free, all samples bore at least two taxa of fungi. Most of the fungi were xerophilic (i.e. grew on media containing 30%–50% sucrose)." This seems to indicate that live fungi are present on the skin of some but not every person and may be attached to and deposited together with the shed skin scales.

In a more recent study, Molva et al. (2019) isolated bacteria and fungi from dust mite cultures. Subsequently they prepared different rearing media, each one enriched with another one of the microbe species. Significant differences in suitability as a rearing diet, revealed by population growth and attractiveness for mites, were demonstrated.

At this point I want to say a few words about products marketed to reduce the allergen content of items such as carpets, rugs and stuffed furniture, of which the active ingredient is a fungicide, i.e., a chemical that kills fungi. The rationale behind it is that mites depend on fungi for their nutrition so that a fungicide must impair the development of a mite population. However, our understanding of dust mite ecology and of the dynamics of allergen accumulation is still inadequate to predict the effect of a fungicide. What could have been done is a field trial, to see if the product works as you hope it will. But this should be done before the product is released on the market, not afterward. Apparently the law does not protect allergy sufferers against such purely commercial enterprises.

9.4 Could humidity influence the food supply?

In 1973 my compatriot professor van Bronswijk wrote, in an article about the population dynamics of dust mites: "… humidity seems to be the most important limiting factor, but it is also possible that humidity affects primarily the growth of microorganisms that are important in the food chain" (van Bronswijk, 1973). She was among the first, perhaps the very first, to propose the possibility that, besides a direct effect of humidity on the development of dust mite populations, there may be an indirect effect as well. This hypothesis hinges on another hypothesis, namely that microorganisms in the house dust ecosystem are perhaps vitally important for house dust mite nutrition. The research group led by Annelies van Bronswijk published several studies on the nutritional value of fungi for dust mites (van de Lustgraaf, 1978a, b; de Saint George-Gridelet, 1987). In these in vitro studies some evidence was found that fungi may supply important nutrients for dust mites, but the ecological impact for the house dust habitat remains obscure.

Hart et al. (2007) studied the development of *D. pteronyssinus* fed on culture medium ("lab food") in comparison with a more natural food (dust enriched with dander). In the Materials and Methods section of their paper they write: "A stock of mattress dust was collected from the beds of a total of 20 nonsmokers and the dust was pooled. It was then frozen at—20°C for a minimum of 1 wk to kill any mites, sieved through a 500-μm mesh, and then kept at room temperature for at least 1 mo before use in experiments. The aim of the latter step was to enable recovery of house dust fungi after the freezing step." So also these authors consider the possibility that fungi might somehow transform and improve the natural food supply for dust mites. A pity that it was not checked whether the one-month incubation indeed makes a difference. The most remarkable finding of this study was the very poor reproduction and survival, in fact failure to survive, of mites that were fed on the "lab diet" at 64%RH. In contrast, at 75%RH the mites thrived on both the lab diet and the enriched dust diet. When fed on the enriched dust diet the mites thrived both at 75%RH and 64%RH (all at 25°C). Only the combination of 64%RH with the lab diet turned out to be dire. I think we need a lot more research before we can explain this. Perhaps then we can also explain the absence of mite populations in certain homes and in public places.

Spores of fungi are resistant to desiccation. But once they germinate, they become vulnerable. It is conceivable that a certain pattern of humidity changes can be tolerated and survived by dust mites but not by fungi.

References

Colloff, M.D., 2009. Dust Mites. CSIRO/Springer Science, ISBN: 978-90-481-2223-3. 583 pp.

de Boer, R., 1990. The control of house dust mite allergens in rugs. J. Allergy Clin. Immunol. 86, 808–814.

de Saint George-Gridelet, D., 1987. Destruction of eggs of *Dermatophagoides pteronyssinus* (Acari: Pyroglyphidae) by Natamycin and Imidazoles in vitro. Int. J. Acarol. 13, 5–14.

Douglas, A.E., Hart, B.J., 1989. The significance of the fungus *Aspergillus peniciloides* to the house dust mite *Dermatophagoides pteronyssinus*. Symbiosis 7, 105–116.

Goldschmidt, H., Kligman, A.M., 1963. Quantitative estimation of keratin production by the epidermis. Arch. Dermatol. 88, 709–712.

Hart, B.J., Le Merdy, L., 1987. Mite allergy, a world wide problem. In: Human Dander-Free House Dust Mite Extracts, Bad Kreuznach, 1–2 September 1987, pp. 47–49.

Hart, B.J., Crowther, D., Wilkinson, T., Biddulph, P., Ucci, M., Pretlove, S., Ridley, I., Orezszyn, T., 2007. Reproduction and development of laboratory and wild house dust mites (Acari: Pyroglyphidae) and their relationship to the natural dust ecosystem. J. Med. Entomol. 44, 568–574.

Hay, D.B., Hart, B.J., Douglas, A.E., 1993. Effects of the fungus *Aspergillus penicillioides* on the house dust mite *Dermatophagoides pteronyssinus*: an experimental re-evaluation. Med. Vet. Entomol. 7, 271–276.

Kligman, A.M., 1964. The biology of the stratum corneum. In: Montagna, M., Lobitz, W.C. (Eds.), The Epidermis. Academic Press, N.Y., London, pp. 395–403.

Molva, V., Nesvorna, M., Hubert, J., 2019. Feeding interactions between microorganisms and the house dust mite *Dermatophagoides pteronyssinus* and *Dermatophagoides farinae* (Astigmata: Pyroglyphidae). J. Med. Entomol. 56, 1669–1677.

Sinha, R.N., van Bronswijk, J.E.M.H., Wallace, H.A.H., 1970. House dust allergy, mites and their fungal associates. C.M.A. J. 103, 300–301.

Spieksma, F.T.M., 1967. The House-Dust Mite *Dermatophagoides pteronyssinus* (Trouessart, 1897), Producer of the House-Dust Allergen (Academic thesis). Rijks Universiteit Leiden, Leiden.

van Bronswijk, J.E.M.H., 1973. *Dermatophagoides pteronyssinus* (Trouessart, 1897) in mattresses and floor dust in a temperate climate. J. Med. Entomol. 10, 63–70.

van Bronswijk, J.E.M.H., Sinha, R.N., 1973. Role of fungi in survival of Dermatophagoides in house-dust environment. Environ. Entomol. 2, 142–145.

van de Lustgraaf, B., 1978a. Seasonal abundance of xerophilic fungi and house-dust mites (Acarida: Pyroglyphidae) in mattress dust. Oecologia 36, 81–91.

van de Lustgraaf, B., 1978b. Ecological relationship between xerophilic fungi and house dust mites (Acarida: Pyroglyphidae). Oecologia 33, 351–359.

van de Lustgraaf, B., Jorde, W., 1977. Pyroglyphid mites, xerophilic fungi and allergenic activity in dust from hospital mattresses. Acta Allergol. 32, 406–412.

CHAPTER TEN

Investigations waiting to be done

I did several suggestions for experiments and investigations that would throw light on yet unclarified aspects of dust mite ecology. The houses and offices that are devoid of mite populations deserve much more attention (Sections 5.1 and 5.4). But also questions of more theoretical interest were addressed. I argued that it must be possible to identify the hypothetical hygroscopic substance or substances that the mites deploy to extract water vapor from the air (Section 4.5.2). And in order to explain the age structure of dust mite populations it would be desirable to have life and fertility tables of mites that live on a lean diet (Sections 7.5, 8.2, and 9.2). In connection with dispersal, sex ratio, and water balance, it would be interesting to examine whether and how mites move directionally in response to a gradient of temperature and RH (Sections 4.2, 7.3, and 7.4). Long established laboratory cultures were used to study the ability of mites to survive adverse physical conditions. However, wild populations must have a much greater genetic variability. Hence their ability to adapt to unfavorable conditions will be greater. It would be interesting to examine whether strains of the same species can be different concerning their ability to withstand extreme conditions. And what about their response to selection? If, for instance, you pick the last surviving mites to propagate a culture after subjecting it to a cold treatment; will that give rise to a more cold resistant strain (Section 6.3)?

But there is one experiment that I have been particularly keen to perform. I never got a chance. Perhaps you do. The idea was spurred by the report of an investigation in Denver (Colorado, USA, Moyer et al., 1985). At high elevations, where houses are usually devoid of dust mites, a mite population survived for 2 years in old furniture that was imported from a location in the valley. This observation may be explained by postulating that food, and not air humidity, was the decisive factor for the survival of these mite populations. It was an incidental observation (see Section 5.3). It could be checked by intentionally bringing old, mite-infested materials into homes at high altitudes. Equally interesting would it be to test whether old materials from homes at high altitudes contain a food source for dust mites. Can mites be reared on fragments of the carpets, rugs, or stuffings?

House Dust Mites
https://doi.org/10.1016/B978-0-443-19111-4.00002-2

The procedures described in Section 9.2 can be followed to investigate this. Alternatively, vacuumed dust could be introduced into new carpet material or into plastic foam and examined for their suitability as a mite habitat (see Fig. 10.1). Both methods have their pros and cons. If these materials do not contain a suitable food source from the beginning, perhaps this will change after some time, when they are exposed to a moist atmosphere. Perhaps inoculation with fungal spores makes a difference? A challenge for experimentalists.

Not only houses at high elevations are interesting subjects. Other buildings with unexplained absence of mites are equally interesting, in particular utility buildings. Here perhaps the concentration of the food may be important. Does the dust vacuumed from a large surface contain enough food if it is introduced into a small piece of carpet?

I would be excited and delighted if my suggestions would inspire somebody to perform such investigations.

Amen

Reference

Moyer, D.B., Nelson, H.S., Arlian, L.G., 1985. House dust mites in Colorado. Ann. Allergy 55, 680–682.

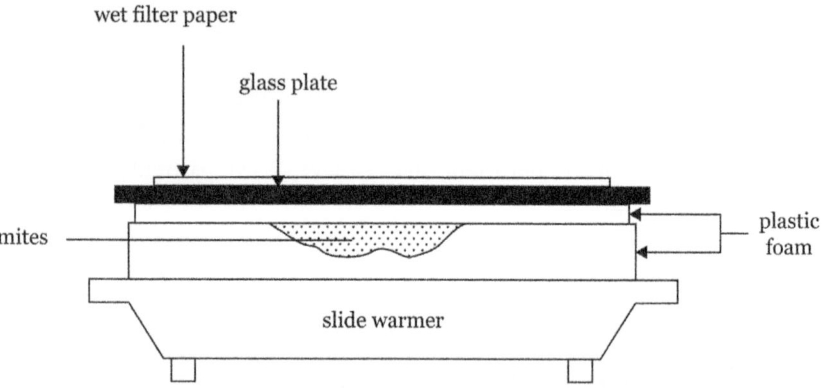

Fig. 10.1 Dust mites can be reared not only in rearing trays but also in plastic foam. Rearing medium of fine dust sinks into the foam and mites live in the pores. Heat from the slide warmer will induce many mites to walk on the glass plate where they can be collected alive with a fine brush. However, some mite species, in particular *Dermatophagoides pteronyssinus*, are reluctant to walk on glass. To capture them the glass plate can be replaced by a sticky tape. Then the stuck mites can be counted but determination of species and live stage will not be possible.

Appendix A

The benefits of taking three samples instead of just one

A.1 Linear regression and the correlation coefficient

Linear regression is a useful statistical technique because it furnishes a possibility for extrapolation. An example will clarify:

Suppose we have vacuumed one and the same rug three times in a row, each time for exactly two minutes. Suppose the amount of dust retrieved by the 1st, 2nd, and 3rd vacuuming is 5.1, 3.9, and 3.3, respectively.

Next we determine the logarithms of these three weights. For reasons that will be outlined later, we suspect that these values decrease in a linear fashion.

So we have:

$x =$ sequential number	1	2	3
$y = \mathbf{Ln}$ (weight of dust)	1.629	1.361	1.194

The best fitting straight line has the formula:

$$y = a + bx$$

with:

$$b = \{\Sigma xy - \Sigma x \Sigma y/n\}/\{\Sigma x^2 - (\Sigma x)^2/n\} \text{ (regression coefficient)}$$

$$a = (\Sigma y - a\Sigma x)/n \quad (y\text{-intercept})$$

$$r = (\Sigma xy - \Sigma x \Sigma y/n)/\{(\Sigma x^2 - (\Sigma x)^2/n)(\Sigma y^2 - (\Sigma y)^2/n)\}^{0.5}$$

$$\text{(correlation coefficient)}$$

$$r^2 = \text{(coefficient of determination)}$$

For the example the values are as follows:

$$b = -0.21750, \quad a = 1.82967, \quad r = -0.99113, \quad \text{and} \quad r^2 = 0.98235$$

The value of r^2 lies between 0 and 1. This value is interesting because it indicates how well the three observed values fit on the straight line. If the three values are exactly on the straight line, then r^2 equals 1. Deviations from the best fitting straight line make r^2 smaller.

A.2 Application

The usual procedure of taking a single sample from a carpet or a rug has two shortcomings. Firstly, it gives you no information of how much dust or how many mites remain in the carpet. Thus there is no way of knowing what the total number of mites per square meter is. Secondly, eggs and immobile stages may be less efficiently captured by vacuuming than the mobile mites. That will result in a false picture of the population structure. Both problems can be overcome by taking three (or more) samples consecutively from the same object. It is essential that these samples are taken from the same area, with equal intensity and equal duration. The total number of mites/m^2 is then estimated through extrapolation. The extrapolation is based on the assumption that the probability for a mite to be captured remains equal through time. Then:

n (time sequence)	Mites present at the start	A_n (captured mites)
1	N	Np
2	$N(1-p)$	$N(1-p)p$
3	$N(1-p)^2$	$N(1-p)^2 p$
4	$N(1-p)^3$	$N(1-p)^3 p$
n	$N(1-p)^{n-1}$	$N(1-p)^{n-1} p$

N = total number of mites originally present in the object

n = sequence of samplings

p = fraction of the number of mites, present at the start of the sampling, that is captured during that sampling

Thus:

$$A_n = N(1-p)^{n-1} p \tag{A.1}$$

The general validity of this relationship can be shown by complete ascertainment.

From formula (A.1) it follows that:

$$\mathbf{Ln}A_n = n\mathbf{Ln}(1-p) + \mathbf{Ln}[Np/(1-p)] \tag{A.2}$$

Thus there is a linear relationship between n and $\mathbf{Ln}A_n$. Whenever we have made three consecutive observations of A_n ($n = 1, 2$, and 3), regression analysis will yield an estimate of the parameters $\mathbf{Ln}(1-p)$ (regression coefficient) and of $\mathbf{Ln}[Np/(1-p)]$ from which the value of p and N can be calculated. Thus we obtain an estimate of N, which is the total number of mites originally present in the sampled object.

Example A1. (from: de Boer and Kuller, 1995)

The dust mite population of a fixed, wall to wall, carpet ($>10\,$yrs) of the living room of a house in Haarlem (the Netherlands) was sampled on 18 March 1994. The carpet was sacrificed and 14 fragments ($13\,cm \times 5\,cm$) were cut from it. The heat escape method was applied three times for one hour to each fragment. The pooled numbers of mites (A_n) from the 14 fragments were as follows:

n	A_n	LnA_n
1	463	6.138
2	143	4.963
3	104	4.644

Regression analysis yields:

$$\mathbf{Ln}A_n = -0.747n + 6.742 \text{ and } r^2 = 0.901$$

So that (see formula A.2)	$\mathbf{Ln}(1-p) = -0.747$	yielding:	$p = 0.526$
and	$\mathbf{Ln}[Np/(1-p)] = 6.742$	yielding:	$N = 763$

The total number of mites actually captured from the 14 fragments was 710, so that on average 50.7 mites were captured per fragment. The variance of these 14 values was much greater, namely 3475.6, which means the mites were very unevenly distributed over the carpet. The calculated number of mites per m^2, 8385, is therefore no better than a rough guess.

With this "heat escape" sampling method, it is not possible to determine species and life stages of the mites. Immobile stages and eggs are not captured at all. But it must be possible to take three consecutively vacuumed samples and follow the same procedure to obtain an estimate for each life stage per species separately, by extrapolation. That will yield an interesting picture of the fauna and population structures (see also Appendix H). To the best of my knowledge, no such study has ever been done.

Reference

de Boer, R., Kuller, K., 1995. Winter survival of house dust mites (*Dermatophagoides* spp.) on the ground floor of Dutch houses. Proc. Exp. Appl. Entomol. N.E.V. 6, 47–52.

Appendix B

Table APDX B Mite species found in house dust from 55 locations, arranged by increasing latitude (i.e., distance from the equator in degrees).

Nr.		x
	Pyroglyphydae	
1	*Pyroglyphidae* sp.	
2	*Dermatophagoides* sp.	
3	*Dermatophagoides pteronyssinus*	55x
4	*Dermatophagoides farinae*	47x
5	*Dermatophagoides microceras*	2x
6	*Dermatophagoides evensi*	2x
7	*Euroglyphus maynei*	41x
8	*Hirstia* sp.	1x
9	*Hirstia simplex*	1x
10	*Hirstia domicola*	4x
11	*Malayoglyphus* sp.	1x
12	*Malayoglyphus intermedius*	6x
13	*Pyroglyphus africanus*	1x
14	*Sturnophagoides* sp.	1x
15	*Sturnophagoides brasiliensis*	1x
	Glycyphagidae	
16	*Glycyphagidae* sp.	8x
17	*Glycyphagus* sp.	9x
18	*Glycyphagus domesticus*	12x
19	*Glycyphagus destructor*	6x
20	*Glycyphagus privatus*	4x
21	*Blomia* sp.	8x
22	*Blomia tropicalis*	7x
23	*Blomia freemani*	1x
24	*Austroglycyphagus geniculatus*	1x
25	*Austroglycyphagus lukoschusi*	1x
26	*Gohieria* sp.	2x
27	*Gohieria fusca*	4x
28	*Lepidoglyphus* sp.	5x
29	*Lepidoglyphus destructor*	6x
30	*Xenoryetes* sp.	1x
	Acaridae	
31	*Acaridae* sp. (undetermined)	2x
32	*Acarus* sp.	4x
33	*Acarus siro*	10x
34	*Acarus farris*	2x

Continued

Table APDX B Mite species found in house dust from 55 locations, arranged by increasing latitude (i.e., distance from the equator in degrees)—cont'd

Nr.		x
35	*Acarus immobilis*	2x
36	*Acarus nidicolous*	1x
37	*Acarus gracilis*	1x
38	*Aleuroglyphus* sp.	1x
39	*Aleuroglyphus ovatus*	1x
40	*Caloglyphus* sp.	2x
41	*Caloglyphus berlesi*	1x
42	*Rhizoglyphus* sp.	2x
43	*Thyreophagus*	2x
44	*Tyrophagus* sp.	10x
45	*Tyrophagus putrescentiae*	13x
46	*Tyrophagus longior*	1x
47	*Tyrophagus palmarum*	1x
	Cheyletidae	
48	*Cheyletidae* sp.	29x
49	*Cheyletus* sp.	11x
50	*Cheyletus eruditus*	9x
51	*Cheyletus trouessarti*	4x
52	*Cheyletus malaccensis*	5x
53	*Cheyletus fortis*	2x
54	*Cheyletia* sp.	1x
55	*Cheyletia fabellifera*	1x
56	*Cheyletomorpha lepidopterorum*	3x
7	*Chelacaropsis moori*	2x
58	*Chelacalopsis* sp.	1x
59	*Cheletomorpha* sp.	1x
60	*Eucheyletia reticulata*	2x
61	*Grallacheles bakeri*	1x
62	*Ker bakeri*	1x
63	*Acaropsis docta*	1x

Each species is designated by the number in the first column (Nr.).

The number is used to indicate the occurrence of the species (Tables APDX B1 and B2, columns 7 and 8).

x: number of locations where the species was found.

Table APDX B1 Pyroglyphidae and Glyciphagidae.

1	2	3	4	5	6	7	8
Author(s)	Year	Geographic location	Lat.	f/m/b	N	Pyroglyphidae	Glycyphagidae
Charlet et al.	1979	Colombia, Cali	3	b	5	3,4,7,12	16
Charlet et al.	1979	Colombia, Saladito	3	b	5	3,4,10,12	16
Charlet et al.	1979	Colombia, Silvia	3	b	5	3,4,7	16
Charlet et al.	1979	Colombia, Fusagasuga (1600 m)	4	b		3,4,7	
Charlet et al.	1979	Colombia, Girardot (300 m)	4	b	5	3,4,7	
Charlet et al.	1979	Colombia, Buenaventura	4	b	5	3,4,7,12,15	16
Charlet et al.	1979	Colombia, Dagua	4	b	5	3,4,7,12	16
Charlet et al.	1979	Colombia, Bogota (2600 m)	5	b		3,4,7,8,11,14	
Hurtado and Parini	1987	Venezuela, Caracas	11	m	54	3,4,7	22
Buchanan and Jones	1972	Zambia, Ndola	13	b	18	3,4	17
Pearson and Cunnington	1973	Barbados	13	b	23	3,4	17,21,30
Malainual et al.	1995	Thailand, Central, North and Northeast	15	m	630	3,4,7	16
Rosa and Flechtman	1979	Brazil, 9 locations	20	b	60	3,4,7,13	22,25,28
Chang and Hsieh	1989	Taiwan, Taipei	25	b	70	3,4,7,12	17,21
Arlian et al.	1992	USA, Delray Beach (Fla.)	26	b	8	3,4,7	22
Sanchez-Covisa et al.	1999	Tenerife (Spain), Santa Cruz and Laguna	28	b	120	3,4,7	18,21,29
Arlian et al.	1992	USA, Galveston (Texas)	29	b	32	3,4,7	22
Arlian et al.	1992	USA, New Orleans (La.)	30	b	58	3,4,7	22
Feldman-Muhsam et al.	1985	Israel, Beer Sheva	31	b	2	3,4,7	21

Continued

Table APDX B1 Pyroglyphidae and Glyciphagidae—cont'd

1 Author(s)	2 Year	3 Geographic location	4 Lat.	5 f/m/b	6 N	7 Pyroglyphidae	8 Glycyphagidae
Colloff et al.	1991	Australia, Perth	32	b	10	3,7	28,29
Feldman–Muhsam et al.	1985	Israel, Bat Yam	32	b	2	3,4,7	21
Feldman–Muhsam et al.	1985	Israel, Jerusalem	32	b	2	3,4,7	21
Feldman–Muhsam et al.	1985	Israel, Nes Ziyyona	32	b	2	3,4,7	21
Ordman	1971	South Africa, 67 towns	32	b	242	3,4,7	
Arlian et al.	1992	USA, San Diego (Calif.)	33	b	25	3,4,7	22
Colloff et al.	1991	Australia, Bunbury	33	b	10	3,7	24,28,29
Feldman–Muhsam et al.	1985	Israel, Nir David	33	b	2	3,4,7	21
Arlian et al.	1992	USA, Los Angeles (Calif.)	34	b	13	3,4	
Lang and Mulla	1977	USA, Coast, California	34	b	15	3,4,6	
Lang and Mulla	1977	USA, Indio, California	34	b	15	3,4	
Lang and Mulla	1977	USA, Lake Arrowhead, California	34	b	15	3,4	
Lang and Mulla	1977	USA, Riverside, California	34	b	15	3,4,6	
Arlian et al.	1992	USA, Memphis (Tenn.)	35	b	31	3,4,7	22
Sakaki and Suto	1995	Japan, Nagoya	35		21	3,4,9	16
Suto et al.	1992a	Japan, Nagoya, wooden houses	35	f	22	3,4,7,10	18,19,20
Suto et al.	1992b	Japan, Nagoya, apartments	35	b	20	3,4,7,10	16,18,19,20
Takaoka and Okada	1984	Japan, Saitama prefecture	35	b	26	3,4,7,10,12	17,26

Author	Year	Location	Lat	f/m/b	N		Refs
Arlian et al.	1992	USA, Greenville (N.C.)	36	b	36	3,4	
Arlian et al.	1992	USA, Cincinnati (Ohio)	39	b	48	3,4	
Goracci et al.	1984	Italy	42	m	37	3,4	
Bigliocchi and Maroli	1995	Italy, Rome	42	b	90	3, 4, 7	18,19,20
Agratorres et al.	1999	Spain, Santiago de C.	43	b	15	3,4,7	18,29
Mumcuoglu	1976	Switzerland, Basel	48	b	190	3,4,7	17,18,19,20,23,26,27
Murray and Zuk	1979	Canada, Vancouver	49	m	3	3,4	
Dusbabek	1975	Czechoslovakia, Prague	50	m	4	3,4	18
Hewitt et al.	1973	UK, Cornwall	50	b	?	3,4,7	17,27
Carswell	1982	UK, Bristol	51	m	51	3,7	
Hart and Whitehead	1990	UK, Oxfordshire	52	b	30	3,7	18
Walshaw and Evens	1987	UK, Liverpool	53	b	50	3,4,7	17,28
Colloff	1987	UK, Glasgow	56	b	74	3,7	17,18,19,27
Haarlov and Alani	1970	Denmark, Copenhagen	56	m	42	3,4,7	20
Sesay and Dobson	1972	UK, Glasgow	56	b	60	3,7	18,19,27
Aas and Mehl	1996	Norway	60	b		3,5,7	18,28,29
Mehl	1998	Norway, 7 locations	60	b		3,5,7	18,29
Stenius and Cunnington	1972	Southern Finland	60	m	37	3,4,7	17,18,19
Stenius and Cunnington	1972	Eastern Finland (Ilomantsi and Uukuniemi)	63	b	8	3,4,7	

Lat.: latitude (distance from the equator in degrees).

f/m/b: f (loor)/m (attress)/b (oth floor and mattress).

N = sample size (number of rooms/mattresses).

Table APDX B2 Acaridae and Cheyletidae.

1 Author(s)	2 Year	3 Geographic location	4 Lat.	5 f/m/b	6 N	7 Acaridae	8 Cheyletidae
Charlet et al.	1979	Colombia, Cali	3	b	5		48
Charlet et al.	1979	Colombia, Saladito	3	b	5		48
Charlet et al.	1979	Colombia, Silvia	3	b	5		48
Charlet et al.	1979	Colombia, Fusagasuga (1600 m)	4	b			48
Charlet et al.	1979	Colombia, Girardot (300 m)	4	b	5		48
Charlet et al.	1979	Colombia, Buenaventura	4	b	5		48
Charlet et al.	1979	Colombia, Dagua	4	b	5		48
Charlet et al.	1979	Colombia, Bogota (2600 m)	5	b			48
Hurtado and Parini	1987	Venezuela, Caracas	11	m	54	45	49
Buchanan and Jones	1972	Zambia, Ndola	13	b	18		49
Pearson and Cunnington	1973	Barbados	13	b	23	33,38,45	48,49,54
Malainual et al.	1995	Thailand, Central, North and Northeast	15	m	630	31	48
Rosa and Flechtman	1979	Brazil, 9 locations	20	b	60	44,45	48,52,61,62
Chang and Hsieh	1989	Taiwan, Taipei	25	b	70	33,44	49
Arlian et al.	1992	USA, Delray Beach (Fla.)	26	b	8		
Sanchez–Covisa et al.	1999	Tenerife (Spain), Santa Cruz and Laguna	28	b	120	41,45	
Arlian et al.	1992	USA, Galveston (Texas)	29	b	32		
Arlian et al.	1992	USA, New Orleans (La.)	30	b	58		
Feldman–Muhsam et al.	1985	Israel, Beer Sheva	31	b	2		48
Colloff et al.	1991	Australia, Perth	32	b	10	32	48,50

Feldman–Muhsam et al.	1985	Israel, Bat Yam	32	b	2		48
Feldman–Muhsam et al.	1985	Israel, Jerusalem	32	b	2		48
Feldman–Muhsam et al.	1985	Israel, Nes Ziyyona	32	b	2		48
Ordman et al.	1971	South Africa, 67 towns	32	b	242		
Arlian et al.	1992	USA, San Diego (Calif.)	33	b	25		
Colloff et al.	1991	Australia, Bunbury	33	b	10	32,37,44	48,50,51,52
Feldman–Muhsam et al.	1985	Israel, Nir David	33	b	2		48
Arlian et al.	1992	USA, Los Angeles (Calif.)	34	b	13		
Lang and Mulla	1977	USA, Coast, California	34	b	15		
Lang and Mulla	1977	USA, Indio, California	34	b	15		
Lang and Mulla	1977	USA, Lake Arrowhead, California	34	b	15		
Lang and Mulla	1977	USA, Riverside, California	34	b	15		
Arlian et al.	1992	USA, Memphis (Tenn.)	35	b	31		
Sakaki and Suto	1995	Japan, Nagoya	35		21	31	48
Suto et al.	1992a	Japan, Nagoya, wooden houses	35	f	22	33,39,42,43,45	48,50,52,53,56,57,60
Suto et al.	1992b	Japan, Nagoya, apartments	35	b	20	33,42,43,45	50,52,53,56,57,60
Takaoka and Okada	1984	japan, Saitama prefecture	35	b	26	45	48,49,58,59
Arlian et al.	1992	USA, Greenville (N.C.)	36	b	36		
Arlian et al.	1992	USA, Cincinnati (Ohio)	39	b	48		
Goracci et al.	1984	Italy	42	m	37		48
Bigliocchi and Maroli	1995	Italy, Rome	42	b	90		

Continued

Table APDX B2 Acaridae and Cheyletidae—cont'd

1	2	3	4	5	6	7	8
Author(s)	Year	Geographic location	Lat.	f/m/b	N	Acaridae	Cheyletidae
Agratorres et al.	1999	Spain, Santiago de C.	43	b	15	45	48,49
Mumcuoglu	1976	Switzerland, Basel	48	b	190	33,34,35,40,44,45,47	49,50,51,55,56
Murray and Zuk	1979	Canada, Vancouver	49	m	3		
Dusbabek	1975	Czechoslovakia, Prague	50	m	4	34,35,45	48,50
Hewitt et al.	1973	UK, Cornwall	50	b	not clear	33,44	48
Carswell et al.	1982	UK, Bristol	51	m	51		48
Hart and Whitehead	1990	UK, Oxfordshire	52	b	30	33,40,45	50,52
Walshaw and Evens	1987	UK, Liverpool	53	b	50		49
Colloff	1987	UK, Glasgow	56	b	74	33,36,43,45,46	48,50,51
Haarlov and Alani	1970	Denmark, Copenhagen	56	m	42	32,33	50,51,63
Sesay and Dobson	1972	UK, Glasgow	56	b	60	33,44	48
Aas and Mehl	1996	Norway	60	b		44,45	49
Mehl	1998	Norway, 7 locations	60	b		44,45	48,49
Stenius and Cunnington	1972	southern Finland	60	m	37	32,44	49
Stenius and Cunnington	1972	eastern Finland (Ilomantsi and Uukuniemi)	63	b	8		

Lat.: latitude (distance from the equator in degrees).

f/m/b: f (loor)/m (attress)/b (oth floor and mattress).

N = sample size (number of rooms/mattresses).

References

Aas, K., Mehl, R., 1996. Pavisning og kvantitering av husstovmidd. Fagbladet Allergi i Praksis 4, 37–40.

Agratorres, J.M., Pareira-Lorenzo, A., Fernandez-Fernandez, I., 1999. Population dynamics of house dust mites in Santiago de Compostela (Galicia, Spain). Acarologia 40, 59–63.

Arlian, L.G., Bernstein, D., Bernstein, I.L., Friedman, S., Grant, A., Lieberman, P., Lopez, M., Mezger, J., Platts-Mills, T., Schatz, M., Spector, S., Wasserman, S.I., Zeiger, R.S., 1992. Prevalence of dust mites in the homes of people with asthma living in eight different geographic areas of the U.S. J. Allergy Clin. Immunol. 90, 292–300.

Bigliocchi, F., Maroli, M., 1995. Distribution and abundance of house dust mites (Acarina: Pyroglyphidae) in Rome, Italy. Aerobiologia 11, 35–40.

Buchanan, D.J., Jones, I.G., 1972. Allergy to house dust mites in the tropics. Br. Med. J. 21, 764.

Carswell, F., Robinson, D.W., Oliver, J., Clark, J., Robinson, P., Wadsworth, J., 1982. House dust mites in Bristol. Clin. Allergy 12, 533–545.

Chang, Y.-C., Hsieh, K.-H., 1989. The study of house dust mites in Taiwan. Ann. Allergy 62, 101–106.

Charlet, L.D., Mula, M.S., Sanchez-Medina, M., Rayes, M.A., 1979. Species composition and population trends of mites in various climatic zones of Colombia. J. Asthma Res. 16, 131–148.

Colloff, M.D., 1987. Mites from house dust in Glasgow. Med. Vet. Entomol. 1, 163–168.

Colloff, M.D., Stewart, G.A., Thompson, P.J., 1991. House dust acarofauna and Der p I equivalent in Australia: The relative importance of Dermatophagoides pteronyssinus and Euroglyphus maynei. Clin. Exp. Allergy 21, 225–230.

Dusbabek, F., 1975. Population structure and dynamics of the house dust mite Dermatophagoides farinae in Czechoslovakia. Folia Parasitologica 22, 219–231.

Feldman-Musham, B., Mumcuoglu, Y., Osterovich, T., 1985. A survey of house dust mites (Acari: Pyroglyphidae and Cheyletidae) in Israel. J. Med. Entomol. 22, 663–669.

Goracci, E., Lazzeri, S., Magnisi, M.T., Sodini, M.L., Duranti, A., 1984. Gli allergo-acari della polvere dei materassi della citta di Livorno. Folia Allergol. Immunol. Clin. 31, 415–421.

Haarlov, N., Alani, M., 1970. House-dust mites (Dermatophagoides pteronyssinus (Trt.), D. farinae Hughes, Euroglyphus maynei (Cooreman) Fain) in Denmark (Acarina). Ent. Scand. 1, 301–306.

Hart, B.J., Whitehead, L., 1990. Ecology of house dust mites in Oxfordshire. Clin. Exp. Allergy 20, 203–209.

Hewitt, M., Barrow, G.I., Miller, D.C., Turk, F., Turk, S., 1973. Mites in the personal environment and their role in skin disorders. Br. J. Dermatol. 89, 401–409.

Hurtado, I., Parini, M., 1987. House dust mites in Caracas. Venezuela. Ann. Allergy 59, 128–130.

Lang, J.D., Mulla, M.S., 1977. Distribution and abundance of house dust mites, Dermatophagoides spp., in different climatic Zones of Southern California. Environ. Entomol. 6, 213–216.

Malainual, N., Vichyanond, P., Phan-Urai, P., 1995. House dust mite fauna in Thailand. Clin. Exp. Allergy 25, 554–560.

Mehl, R., 1998. Occurrence of mites in Norway and the rest of Scandinavia. Allergy 53 (Suppl. 48), 28–35.

Mumcuoglu, Y., 1976. House dust mites in Switzerland I. Distribution and taxonomy. J. Med. Entomol. 13, 361–373.

Murray, A.B., Zuk, P., 1979. The seasonal variation in a population of house dust mites in a North American city. J. Allergy Clin. Immunol. 64, 266–269.

Ordman, D., 1971. The incidence of 'climate asthma' en South Africa: its relation to the distribution of mites. S. Afr. Med. J., 739–740.

Pearson, R.S.B., Cunnington, A.M., 1973. The importance of mites in house dust sensitivity in barbadian asthmatics. Clin. Allergy 3, 299–306.

Rosa, A.E., Flechtmann, C.H.W., 1979. Mites in house dust from Brazil. Int. J. Acarol. 5, 195–198.

Sakaki, J., Suto, C., 1995. Cluster analysis of domestic mites and housing conditions in wooden houses in Nagoya, Japan. Jpn. J. San. Zool. 46, 41–48.

Sanchez-Covisa, A., Rodrigues-Rodrigues, J.A., de la Torre, F., Garcia-Robaina, J.C., 1999. Faune d'acarien de la poussiere domestique dans l'ile de Teneri. Acarologia 15, 55–58.

Sesay, H.R., Dobson, R.M., 1972. Studies on the mite fauna of house dust in Scotland with special reference to that of bedding. Acarologia 14, 384–392.

Stenius, B., Cunnington, A.M., 1972. House dust mites and respiratory allergy: a qualitative survey of species occurring in Finnish house dust. Scand. J. Respir. Dis. 53, 338–348.

Suto, C., Sakaki, I., Itoh, H., Mitibata, M., 1992a. Studies on the ecology of house dust mites in wooden houses in Nagoya, with special reference to the influence of room ratios on the prevalence of mites and allergy. Jpn. J. San. Zool. 43, 217–228.

Suto, C., Sakaki, I., Itoh, H., Mitibata, M., 1992b. Influence of floor levels on the prevalence of house-dust mites in apartments in Nagoya, Japan. Jpn. J. San. Zool. 43, 307–318.

Takaoka, M., Okada, S., 1984. Ecological studies on the house dust mites in Saitama prefecture Japan; Seasonal occurrences of the pyroglyphid mites in house dust in 1981 to 1982. Jpn. J. San. Zool. 35, 129–137.

Walshaw, M.J., Evans, C.C., 1987. The effect of seasonal and domestic factors on the distribution of Euroglyphus maynei in the homes of Dermatophagoides pteronyssinus allergic patients. Clin. Allergy 17, 7–14.

Appendix C

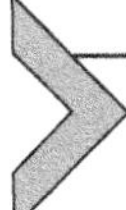

Calculation of absolute humidity (AH) from readings of relative humidity (RH) and temperature (*T*)

There is a rich choice of measures to quantify the humidity of air. Besides relative humidity (%RH) and absolute humidity (AH) we have water activity (aw or av), saturation deficit, dew point, and so on. Each one of them has its pros and cons. For the early studies on dust mite ecology, the group around Wharton and Arlian chose water vapor activity (av) as the preferred measure of humidity. But in later papers by these authors and by everybody else, the more usual %RH was used as a measure of air humidity. As long as we are concerned with only the gaseous phase (air) the conversion is very simple:

$$\mathrm{RH} = \mathrm{av} \times 100\%$$

For every temperature there is a maximum amount of water that can be contained as vapor in a volume of air. But usually only a percentage of this maximum is actually present, hence "percent relative humidity." Since the maximum amount that can be contained in the air depends on the temperature, it is meaningless to record the %RH without also recording the temperature.

Absolute humidity (AH) is the quantity of water in a particular volume of air. Often it is useful, or even necessary, to calculate the absolute humidity (AH) from readings of the %RH and the temperature.

This can be done using the following equations (Buck, 1981):

ews = water vapor partial pressure of saturated moist air

Within the temperature range $-20°C$ to $+55°C$:

$$\mathrm{ews} = 6.1365 \ \exp\left[(17.502\,T)/(240.97 + T)\right] \quad \mathrm{mbar} \qquad (\mathrm{C.1})$$

$T =$ temperature in °C

ew = water vapor partial pressure

$$\mathrm{ew} = (\%\mathrm{RH}/100) \ \mathrm{ews} \quad \mathrm{mbar} \qquad (\mathrm{C.2})$$

ews − ew = saturation deficit

AH = water content of 1 kg of dry air in g water

$$AH = 621.97\, ew/(P - ew) \quad g/kg \qquad (C.3)$$

P = atmospheric pressure in mbar

Approximately:

$$AH \approx 0.614\, ew \quad g/kg$$

Examples:

T (°C)	ews (mbar)	%RH	≈AH (g/kg)
−10	2.88	100	1.77
−10	2.88	50	0.88
0	6.14	100	3.77
0	6.14	50	1.88
10	12.32	100	7.57
10	12.32	50	3.78
20	23.47	100	14.41

Reference

Buck, A.L., 1981. New equations for computing vapor pressure and enhancement factor. J. Appl. Meteorol. 20, 1527–1532.

Appendix D

Determination of the CEH

Preliminary tests indicated that the critical equilibrium humidity of our strain of *Dermatophagoides pteronyssinus* at 20°C was between 56 and 60%RH. A more rigorous test to check this was done as follows: Adult females from the stock culture were partially dehydrated (1 week at <5% RH and 25°C). Then they were weighed and exposed to either test humidity at 20°C during 48 h. Five of nine mites exposed to 60%RH gained weight significantly (closed circles in the figure, t-test, one sided, $P < .001$ for two mites, $P < .01$ for two other mites, $P < .05$ for one mite), showing that the CEH was below 60%. None of seven mites exposed to 56% and none of 10 mites exposed to 58% showed a significant increase of weight (open circles in the figure) confirming that the CEH at 20°C is probably between 58 and 60%RH (i.e., 8.4–8.6 g/kg AH).

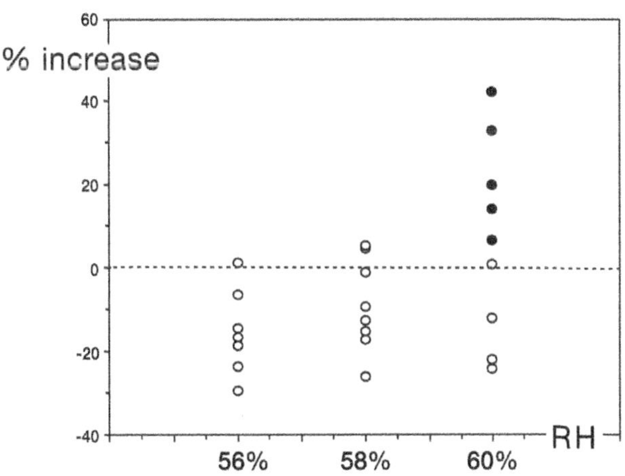

We did an extra check to confirm that the CEH at 20°C is >58%. Subcultures, containing hundreds of mites and eggs, were transferred from standard culturing conditions of 75%RH and 20°C to a constant 58%RH at 20°C. After several weeks a few mites remained alive but were obviously dehydrated. These regained a normal water-logged appearance within 48 hours when transferred to 60%RH without food.

Arlian and Veselica (1981) determined the CEH values of *Dermatophagoides farinae* at four different temperatures. They weighed fasting and partly dehydrated mites individually and subsequently exposed them for 24 h to a certain combination of temperature and RH. If this combination is below the CEH, the mite will continue to lose water, i.e., lose weight. Above the CEH it will gain weight. The weight change in 24 h is plotted against the RH. By linear interpolation, the RH value associated with zero weight change, i.e., the CEH value, is estimated.

Yet another procedure was followed by Knülle (1962), working with the flour mite *Acarus siro*. As a pretreatment, the mites were kept for 16 h in desiccating conditions (70%RH and 22°C) without food. About 10,000 of these mites were confined in a small cage and weighed. Then they were exposed for either 2 or 6 h to a test humidity and weighed again. When exposed to 70%RH, the mites continued to lose weight, but at 75% or more they gained weight so that the CEH at 22°C must be between 70% and 75%RH.

References

Arlian, L.G., Veselica, M.M., 1981. Effect of temperature on the equilibrium body water mass in the mite *Dermatophagoides farinae*. Physiol. Zool. 54, 393–399.
Knülle, W., 1962. Die Abhangigkeit der Luftfeuchte-Reaktionen de Mehlmilbe (*Acarus siro*) vom Wassergehalt des Korpers. Zeitschr. vergl. Physiol. 45, 233–246.

Appendix E

Mites and animal nests

Pyroglyphid mites and the animal species from which, or from the nests of which, they were collected. Bird species were categorized according to nesting behavior as follows (column 4, Category):

1: Humans (house dust and in one case granary dust); $N = 15$.

2: Mammal nest or body; $N = 3$.

3: Bird body; $N = 2$.

3F: "free" bird nest (between branches); $N = 14$.

3H: Bird nest in a tree hole or a crevice; $N = 17$.

3G: Nest of bird with gregarious breeding habit; $N = 8$.

4: Bird nest difficult to categorize.

Mite species	Location	On or in the nest of	Category
Dermatophagoidinae			
Dermatophagoides anisopoda	Af	Agapornis pullaria (Psittacidae)	3H
Dermatophagoides aureliani	Af	Passer griseus ugandae (Ploceidae)	3H
Dermatophagoides evansi	S	bird	3
Dermatophagoides evansi	NA	Iridoprocne bicolor (Hirundinidae)	3H
Dermatophagoides evansi	NA	Passer domesticus (Ploceidae)	3H
Dermatophagoides evansi	SA	Petrochelidon fulva (Hirundinidae)	3G
Dermatophagoides evansi	NA	Quiscalus quiscula (Icteridae)	3F
Dermatophagoides evansi	As	human (house dust)	1
Dermatophagoides farinae	S	human (house dust)	1
Dermatophagoides microceras	S	human (house dust)	1
Dermatophagoides neotropicalis	S	human (house dust)	1
Dermatophagoides pteronyssinus	S	human (house dust)	1
Dermatophagoides rwandae	Af	Buphagus africanus (Buphagidae)	3H

Continued

—cont'd

Mite species	Location	On or in the nest of	Category
Dermatophagoides sclerovestibulatus	Af	Buphagus erythrorhynchus (Buphagidae)	3H
Dermatophagoides siboney	S	human (house dust)	1
Dermatophagoides simplex	SA	Passer domesticus (Ploceidae)	3H
Fainoglyphus magnasternum	SA	Certhiaxis erythropos (Funariidae)	3F
Guatemalichus tachornis	SA	Tachornis phoenicobia iradii (Apodidae)	3G
Hirstia chelidonis = D. passericola	E	Apus apus (Apodidae)	3H
Hirstia chelidonis = D. passericola	E	Delichon urbica (Hirundinidae)	3G
Hirstia chelidonis = D. passericola	E	Delichon urbica (Hirundinidae)	3G
Hirstia chelidonis = D. passericola	E	Passer domesticus (Ploceidae)	3H
Hirstia domicola	S	human (house dust)	1
Malayoglyphus carmelitus	S	human (house dust)	1
Malayoglyphus intermedius	S	human (house dust)	1
Onychalges asaphospathus	Af	Clytospiza monteiri (Estrildidae)	4
Onychalges asaphospathus	Af	Euschistospiza dybovskyi (Estrildidae)	4
Onychalges asaphospathus	Af	Spermestes bicolor (Estrildidae)	4
Onychalges nidicola	Af	Passer domesticus (Ploceidae)	3H
Onychalges odonturus	Af	Spermophaga haematina (Estrildidae)	3F
Onychalges pachyspathus	Af	Estrilda astrild (Estrildidae)	3F
Onychalges pachyspathus	Af	Estrilda atricapilla (Estrildidae)	3F
Onychalges pachyspathus	Af	Estrilda melpoda (Estrildidae)	3F
Onychalges pachyspathus	Af	Estrilda nonnula (Estrildidae)	3F
Onychalges schizurus	Af	Lagonosticta rubricata (Estrildidae)	3F
Onychalges spinitarsis	Af	Pogoniulus scolopaceus (Capitonidae)	3H
Paramealia ovata	Af	Pleucus nigricollis brachypterus (Ploceidae)	3F / 3G
Pottocola (Capitonocoptes) longipennis	Af	Pogoniulus scolopaceus (Capitonidae)	3H
Pottocola (Capitonocoptes) lybius	Af	Lybius dubius (Capitonidae)	3H
Pottocola (Capitonocoptes) lybius	Af	Lybius rubrifacies (Capitonidae)	3H
Pottocola (Capitonocoptes) lybius	Af	Lybius torquatus (Capitonidae)	3H

—cont'd

Mite species	Location	On or in the nest of	Category
Pottocola (Capitonocoptes) *lybius*	Af	Lybius ventriscutata (Capitonidae)	3H
Pottocola (Capitonocoptes) *lybius*	Af	Lybius vielloti (Capitonidae)	3H
Sturnophagoides bakeri	NA	Starling (Sturnidae)	3H
Sturnophagoides bakeri	NA	unidentified birds	3
Sturnophagoides brasiliensis	S	unidentified birds	1
Sturnophagoides petrochelidonis	SA	Petrochelidon fulva (Hirundinidae)	3G
Pyroglyphinae			
Asiopyroglyphus thailandicus	As	Meiglyptes tristis (Picidae)	3H
Bontiella bouilloni	Af	Cinnyris venustus (Nectariniidae)	3F
Bontiella bouilloni	Af	Colius striatus (Coliidae)	3F
Bontiella bouilloni	Af	Grammomys surdaster (Muridae)	2
Bontiella bouilloni	Af	Nectarinia kilimensis (Nectariniidae)	3F
Bontiella bouilloni	Af	Spermestes cucullatus (Ploceidae)	3F
Bontiella bouilloni	Af	Textor cucullatus (Ploceidae)	3G
Campephilocoptes atyeoi	SA	Campephilus rubricollis (Picidae)	3H
Campephilocoptes paraguayensis	SA	Campephilus leucopogon (Picidae)	3H
Euroglyphus maynei	S	human (house dust)	1
Gymnoglyphus longior	S	human (house dust)	1
Gymnoglyphus losu	NA	granary dust	1
Hughesiella africana	As	human (house dust)	1
Hughesiella africana	SA	human (house dust)	1
Hughesiella africana	SA	Passer domesticus (Ploceidae)	3H
Pyroglyphus morlani	NA	Neotoma albigula (Cricetidae)	2
Pyroglyphus morlani	NA	Neotoma spec. (Cricetidae)	2
Weelawadjia australis	Au	guano	guano
Weelawadjia australis	Au	Hirundo neoxena (Hirundinidae)	3G
Weelawadjia australis	Au	Lonchura castaneothorax (Ploceidae)	3F
Paralgopsinae			
Paralgopsis ctenodontus	SA	Ara macao (Psittacidae)	3H
Paralgopsis paradoxus	SA	Pyrrhura leucotis (Psittacidae)	3H

Column 2 (location): *Af*, Africa; *As*, Asia; *Au*, Australia; *E*, Europe; *NA*, North America; *SA*, South and Central America; *S*, several locations.
Based data provided by Fain et al. (1990) and Colloff (2009).

References

Colloff, M.D., 2009. Dust Mites. CSIRO/Springer Science. 583 pp. ISBN 978-90-481-2223-3.

Fain, A., Guerin, B., Hart, B.J., 1990. In: Guerin, B. (Ed.), Mites and Allergic Disease. Allerbio, Varenne en Argonne, France. 190 pp.

Appendix F

Intrinsic rate of increase and population structure

F.1 The mathematical basics

Changes of the size of a population can be written as follows:

$$\partial N/\partial t = B + I - D - E$$

where t is the time; N is the number of individuals; and B, I, D, and E are the rates at which individuals are born, immigrate, die, and emigrate. In a small population, all four of these rates may be significant. In larger populations, immigration and emigration may be relatively unimportant. We consider a large population and set $I = E = 0$. In the simplest model of exponential growth, B and D are each proportional to the number of individuals (N). In other words, $B = bN$ and $D = dN$, where b and d are the average birth and death rates per individual per unit time. Then:

$$\partial N/\partial t = bN - dN = (b - d)N = rN$$

where r $(=b-d)$ is called "the instantaneous coefficient of population growth" and is also referred to as the "intrinsic rate of (natural) increase" or as the "Malthusian parameter."

The solution of the equation is as follows:

$$N(t) = N(0)e^{rt}$$

where $N(0)$ is the number of individuals in the population at the moment we begin our observations and t is the time elapsed after the observations begin.

Theoretically, there is an optimum environment for every population, where its r would reach the maximum possible value "$r(\max)$."

Consider the simple hypothetical "life table" as follows:

x (years)	l(x)	m(x)	l(x)m(x)
0	1	0	0
1	0.5	2	1
2	0.2	4	0.8
3	0		0
Total			$1.8 = \Sigma l(x)m(x) = R_o$

$l(x)$ is the proportion of individuals that survive from birth to age x ($0 \leq l(x) \leq 1$). Of course, the life table is based on observations of a limited number of individuals, so that the value of $l(x)$ is in fact an estimate of the "true" value of it.

$m(x)$ is the number of female offspring per individual female per unit time, i.e., the number of female offspring produced in the time interval $x - 1$ to x, divided by the number of females alive at time x. Again, the life table provides but an estimate of the "true" value.

R_o is called "the net reproductive rate" and is defined as the average number of female offspring produced by each female during her entire lifetime.

Starting from $N(t) = N(0)e^{rt}$

Let $t =$ maximum age that a female can reach and $N(0)$ be only one female. Thus we have set out to find the number of female descendants a single female will produce, including her own daughters, the daughters they have, etc., during the maximum life span one female can enjoy.

Since $N(0) = 1$

$$N(\text{max age}) = e^{r(\text{max age})}$$

$$= \Sigma l(x)m(x)e^{r(\text{max age}-x)}$$

i.e., the total number of female individuals stemming from a single female is the sum of the expected number of daughters produced by that female at each age x of the female ($l(x)m(x)$) times the number of daughters that each of these sets of offspring will produce from the time of their birth to the maximum age of the original female (max age $- x$). Substituting and rearranging we obtain:

$$e^{r(\text{max age})} = e^{r(\text{max age})}\Sigma l(x)m(x)e^{-rx}$$

so that:

$$\Sigma l(x)m(x)e^{-rx} = 1$$

This equation can be used to find the value of r by a trial-and-error procedure, trying so many values of r until one is found that makes $\Sigma l(x)m(x)e^{-rx} \approx 1$.

x	$l(x)m(x)$	$l(x)m(x)e^{-rx}$ $(r = 0.420)$
0	0	0.0000
1	1	0.6570
2	0.8	0.3454
3	0	0.0000
	1.8	**1.0024**
	$=R_o$	$=\Sigma l(x)m(x)e^{-rx}$

r	$\Sigma l(x)m(x)e^{-rx}$
0.400 ►	1.0298
0.500 ►	0.9008
0.450 ►	0.9629
0.420 ►	1.0024

F.2 The intrinsic rate of increase, doubling time, and the stable age (stage) distribution

Life history data of a strain of *Dermatophagoides pteronyssinus* under near optimum physical conditions (23°C and 75%RH) and with food ad libitum were determined by Arlian et al. (1990). The daily egg production and survival fraction data had to be read from a graph. I recalculated the survival and fecundity data for intervals lasting 8 days. Then the first interval corresponds roughly with the egg stage, the second with the larval stage, the third with the protonymph, etc. Arlian's fecundity and longevity data are based on 56 inseminated females that were observed from the tritonymph–adult ecdysis onward. Preadult mortality was 14%. To account for 14% juvenile mortality the starting number is set at 65 female eggs, and the mortality is spread evenly over the four immature life stages, thus arriving at 56 adult females in the fifth period, when egg production commences. It is assumed that half of all eggs are of the female sex (Table F1).

Birch (1948) described a method to compute the value of r and the stable age distribution, from the life and fertility table. He recommends to use some convenient measure of time, such as days or weeks. But we are using a life table where time is measured in 8-day intervals. Following Birch's procedure I arrived at a value of $r = 0.588$ per 8-day interval, per head, which corresponds with $r = 0.0735$ per day, per head. Details are given in Tables F2 and F3. The survivorship function, $l(x)$, in the table is the fraction

Table F1 Life and fertility table of *Dermatophagoides pteronyssinus*.

Stage	Days	x (8 day intervals)	Females (survival)	♀Eggs
	0	0	65	0
Egg	8	1	63	0
Larva	16	2	61	0
Protonymph	24	3	59	0
Tritonymph	32	4	57	0
Adult age class 1	40	5	56	328.0
Adult age class 2	48	6	54	711.0
Adult age class 3	56	7	43	433.0
Adult age class 4	64	8	31	254.0
Adult age class 5	72	9	14	143.5
Adult age class 6	80	10	4	41.5
Adult age class 7	88	11	2	16.5
Adult age class 8	96	12	0	1.0

Based on data published by Arlian, L.G., Rapp, C.M., Ahmed, S.G., 1990. Development of *Dermatophagoides pteronyssinus*. J. Med. Entomol. 27, 1035–1040.

Table F2 Computation of r from the life and fertility tables.

x (pivotal age in 8-day intervals)	$l(x)$	$m(x)$	$l(x)m(x)$	$l(x)m(x)e^{-rx}$ ($r = 0.588$)		
0.5	0.9846	0.0000	0.0000	0.0000		
1.5	0.9538	0.0000	0.0000	0.0000	0.585 ►	1.016500
2.5	0.9231	0.0000	0.0000	0.0000		
3.5	0.8923	0.0000	0.0000	0.0000	**0.588 ►**	**1.000042**
4.5	0.8692	5.8053	5.0462	0.3579		
5.5	0.8462	12.9273	10.9385	0.4310		
6.5	0.7462	8.9278	6.6615	0.1458	0.590 ►	0.989223
7.5	0.5692	6.8649	3.9077	0.0475		
8.5	0.3462	6.3778	2.2077	0.0149	0.600 ►	0.936905
9.5	0.1385	4.6111	0.6385	0.0024		
10.5	0.0462	5.5000	0.2538	0.0005		
11.5	0.0154	1.0000	0.0154	0.0000		
			29.67 $=\Sigma l(x)m(x)$ $=R_0$	**1.000042** $=\Sigma l(x)m(x)e^{-rx}$		

Based on data published by Arlian, L.G., Rapp, C.M., Ahmed, S.G., 1990. Development of *Dermatophagoides pteronyssinus*. J. Med. Entomol. 27, 1035–1040 (see Table F1).

Table F3 Computation of the stable age distribution from the life and fertility tables (see Tables F1 and F2).

Age group (x)	L(x)	$e^{-r(x+1)}$ (r = 0.588)	$L(x)e^{-r(x+1)}$	$100\beta L(x)$ $e^{-r(x+1)}$	$\Sigma 100\beta L(x)e^{-r(x+1)}$	
0 –	0.9846	0.5554	0.547	46.560	46.6	%E
1 –	0.9538	0.3085	0.294	25.053	71.6	%E + L
2 –	0.9231	0.1714	0.158	13.466	85.1	%E + L + P
3 –	0.8923	0.0952	0.085	7.230	92.3	%E + L + P + T
4 –	0.8692	0.0529	0.046	3.912	96.2	
5 –	0.8462	0.0294	0.025	2.115	98.3	
6 –	0.7462	0.0163	0.012	1.036	99.4	
7 –	0.5692	0.0091	0.005	0.439	99.8	
8 –	0.3462	0.0050	0.002	0.148	100.0	
9 –	0.1385	0.0028	0.000	0.033	100.0	
10 –	0.0462	0.0016	0.000	0.006	100.0	
11 –	0.0154	0.0009	0.000	0.001	100.0	%E + L + P + T + A
			$\Sigma = 1.175$ $= 1/\beta$ $\rightarrow \beta = 0.8513$			

E, eggs; *L*, larvae; *P*, protonymphs; *T*, tritonymphs; *A*, adults.

of the 65 females, surviving from birth to age x. The maternity function, $m(x)$, is the number of female eggs produced per 8-day time unit by a female aged x. We have to rely on estimates of the values of $l(x)$ and $m(x)$, based on a limited number of observations (Table F1). The value of r is determined by a trial-and-error procedure making

$$\Sigma e^{-rx}l(x)m(x) \approx 1.$$

Instead of the Malthusian parameter r, the doubling time can be given to indicate the speed of population increase. Once the population has reached a stable age structure and is increasing exponentially, the size of the population is doubled in a fixed period t_d the "doubling time." The value of t_d can be calculated on the basis of the formula $N(t) = N(0)e^{rt}$, when $N(t_d) = 2N(0)$.

After some rearrangements we obtain:

$$t_d = (\text{Ln}2)/r \text{ and } r = (\text{Ln}2)/t_d$$

With $r = 0.0735$ per day, per head, we obtain: $t_d = (\text{Ln}2)/0.0735 = 9.43$ days.

Thus the population will double its size every 9.43 days.

R_o (Table F2) is the so-called net reproductive rate. It expresses the number of times the population multiplies each generation. Thus, in optimum conditions and with unlimited food, a dust mite population will multiply 29.67 times each generation.

T is the "generation time." It is the mean age of the mothers at birth of female offspring and it can be calculated with the formula:

$T = (LnR_o)/r = (Ln29.67)/0.0735 = 46.12$ days. So, if you happen to catch a female in the act of giving birth to a female egg, her average age will be 46.12 days.

The survival and egg production of *Dermatophagoides pteronyssinus* was recorded by Arlian et al. (1990) day by day. I pooled the survival and fertility data of periods lasting 8 days, because a less detailed life table is often desirable, especially for the application of Leslie matrices (see next section). Thus much of the more detailed information is ignored. In order to check whether this has a significant influence on the results, I also calculated r and the stable age distribution on the basis of a more detailed life table. In Table F4, the values of r and the stable age distribution are given when time intervals are only 1 day and compared with the values obtained with 8-day intervals.

There are a few more studies available with life history data of *Dermatophagoides pteronyssinus*. Information on juvenile mortality observed in various studies is given in Table F5. As can be seen, mortality rates are quite different in different studies, perhaps due to strain differences or to experimental circumstances.

Table F4 Calculated stable stage distribution, Malthusian parameter (*r*), doubling time, net reproductive rate (R_0), and generation time (*T*) of *D. pteronyssinus* at 23°C, 76%RH, and plenty of food.

	1-day interval	8-day intervals
% 0–8 days old (eggs)	46.3	46.6
% 9–16 days old (larvae)	25.1	25.0
% 17–24 days old (protonymphs)	13.5	13.5
% 25–32 days old (tritonymphs)	7.3	7.2
% >32 days old (adults)	7.8	7.7
r (per day, per head)	0.0729	0.0735
doubling time (days)	9.51	9.43
R_o $(=\Sigma l(x)m(x))$	29.67	29.67
T (days)	46.50	46.12

Table F5 Preadult mortality of *Dermatophagoides pteronyssinus* in near optimum conditions.

Source	°C/%RH	Stage(s)	Initial number	Number surviving	Mortality (%)
Colloff (1987a)	Alternating	E[a]	35	*35*	0.0
	16 h. 15°/60%	E[b]	35	*34*	2.9
	8 h. 30°/75%	E[c]	35	*31*	11.5
		E[d]	35	*24*	31.5
Colloff (1987b)	25°/75%	E	100	*89*	*11.2*
		E + L	100	*83*	*17.3*
Gamal-Eddin et al. (1983)	25°/75%	E + L + P + T	50	*40*	20.0
Matsumoto et al. (1986)	25°/76%	E	35	28	*20*
		L	28	22	*21.4*
		P	22	21	*4.5*
		T	21	21	*0.0*
		E +L +P + T	35	21	*40.0*
Arlian et al. (1990)	23°/76%	E + L + P +T	175		*14.0*

[a]Wild strain fed on house dust
[b]Wild strain fed on yeast
[c]Laboratory strain fed on house dust
[d]Laboratory strain fed on yeast
In italics: data obtained indirectly from cited sources (either recalculated or read from a graph).

In Table F6, r and the stable age distribution are given when the preadult mortality is set at 10%, 20%, and 40% and compared with the values obtained when preadult mortality is set at 14%. The mortality was spread evenly over the immature stages.

So far, I did not distinguish between active and quiescent mites. In Table F6, the stable age distribution (calculated with an interval length of 1 day) is given while taking these previously ignored details into consideration.

F.3 Leslie matrices and population development

Developments of the population size and structure can also be modeled by matrix multiplications. The relation between the continuous age variable x,

Table F6 The stable stage distribution and r when preadult mortality is either 10%, 14%, 20%, or 40%.

Preadult mortality	10%		14%		20%		40%	
% Eggs	47.1		46.4		46.9		46.9	
% Larvae[a]	29.2	(6.5)	29.3	(9.1)	29.3	(9.1)	29.3	(9.1)
% Protonymphs[a]	10.3	(3.7)	10.5	(3.8)	10.4	(3.8)	10.3	(3.7)
% Tritonymphs[a]	6.6	(2.1)	6.7	(2.0)	6.6	(2.0)	6.4	(1.9)
% Adults	6.8		7.1		6.7		7.2	
% Quiescent mites	12.3		14.9		14.9		14.7	
r (days)	0.0757		0.0729		0.07295		0.0648	
Doubling time (days)	9.16		9.51		9.50		10.70	
R_o	31.1		29.67		27.55		20.74	
T (days)	45.41		46.5		45.56		46.79	

[a]In brackets: % quiescent.

used in the life table functions $m(x)$ and $l(x)$, and the discrete age classes i, used in the projection matrix parameters P_i and F_i, can be shown as follows:

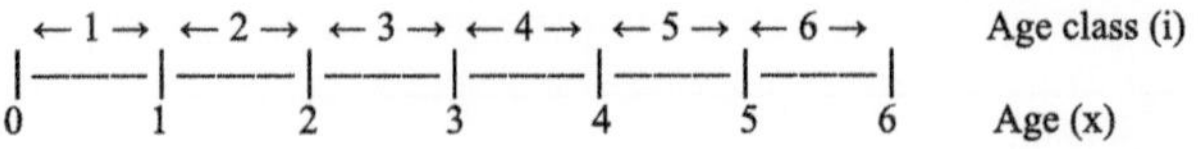

$$\underset{0}{|}\ \overset{\leftarrow 1 \rightarrow}{\underset{1}{---}}\ \overset{\leftarrow 2 \rightarrow}{\underset{2}{---}}\ \overset{\leftarrow 3 \rightarrow}{\underset{3}{---}}\ \overset{\leftarrow 4 \rightarrow}{\underset{4}{---}}\ \overset{\leftarrow 5 \rightarrow}{\underset{5}{---}}\ \overset{\leftarrow 6 \rightarrow}{\underset{6}{---}}|\qquad \text{Age class (i)}$$

Age (x)

The (constant) values of m_i are approximated by dividing the number of female eggs, produced during the interval i, by the median number of females alive during the interval. l_i is the number of females alive at time $x = i$ (the end of the interval) divided by the number alive in the beginning ($x = 0$). The numbers of individuals in each of the 12 age classes at time t can be written in vector form, $n(t)$, and the numbers at the end of the subsequent time interval, $n(t + 1)$, can be obtained by matrix multiplication:

$$
\begin{bmatrix} n_1 \\ n_2 \\ n_3 \\ \cdots \\ n_s \end{bmatrix}(t+1) =
\begin{bmatrix}
F_1 & F_2 & F_3 & \cdots & F_2 \\
P_1 & 0 & 0 & \cdots & 0 \\
0 & P_2 & 0 & \cdots & 0 \\
\cdots & \cdots & \cdots & \cdots & \cdots \\
0 & 0 & \cdots & P_{s-1} & 0
\end{bmatrix}
\begin{bmatrix} n_1 \\ n_2 \\ n_3 \\ \cdots \\ n_s \end{bmatrix}(t)
$$

where: n_i is the number of individuals in age class i

$$P_i \approx [l(i) + l(i + 1)]/[l(i - 1) + l(i)]$$

and

$$F_i = l(0.5)((m_i + P_i m_{i+1})/2)$$

(see: Caswell, 1989)

The value of $l(0.5)$ is not known directly, but is estimated by linear interpolation:

$$l(0.5) \approx [l(0) + l(1)]/2$$

Thus $l(0.5) \approx 0.98$ and the values of m_i, F_i, and P_i are given in Table F7.

Table F7 For explanation, see text.

i	l_i	m_i	P_i	F_i
1	0.9692	0.0000	0.9688	0.0000
2	0.9385	0.0000	0.9677	0.0000
3	0.9077	0.0000	0.9667	0.0000
4	0.8769	0.0000	0.9741	2.7841
5	0.8615	5.8053	0.9735	9.0532
6	0.8308	12.9273	0.8818	10.2400
7	0.6615	8.9278	0.7629	6.9735
8	0.4769	6.8649	0.6081	5.2890
9	0.2154	6.3778	0.4000	4.0479
10	0.0615	4.6111	0.3333	3.1726
11	0.0308	5.5000	0.3333	2.8718
12	0.0000	1.0000	0.0000	0.4923

When the symbol A is used to denote the projection matrix, a matrix multiplication can be written, more compactly, as follows:

$$n(t + 1) = A_n(t).$$

Starting with:

$$
n(0) = \begin{bmatrix} 0 \\ 0 \\ 0 \\ 0 \\ 1 \\ 0 \\ 0 \\ 0 \\ 0 \\ 0 \\ 0 \\ 0 \end{bmatrix}
$$

means that at the start, i.e., at $t = 0$, there is only one young female (adult age class 1). This female is inseminated and starts to reproduce. The composition of the resulting population in subsequent generations is given in Table F8 and depicted in Fig. 8.1 (Chapter 8).

By day 224 (28 8-day periods), the age distribution is practically constant and the size of the population is increasing exponentially according to the formula[a]:

$$
N(t_2) = N(t_1)e^{r(t2-t1)}
$$

After some rearrangements, we obtain:

$$
r = [\mathrm{Ln}(N(t_2)) - \mathrm{Ln}(N(t_1))]/(t_2 - t_1)
$$

which formula can be used to calculate the value of r when t_1, t_2, $N(t_1)$, and $N(t_2)$ are known.

If we look at Table F8 and start, for instance, from the situation on day 224 (i.e., after 28 8-day intervals), i.e., set $t_1 = 224$, then $N(t_1) = 82{,}425{,}641$ mites and eggs. At some later time, for instance on day 384 (48 intervals), $N(t_2) = 13{,}919{,}005{,}330{,}000$, while $t_2 = 384$ days.

The value of r can then be calculated:

$$
r = [\mathrm{Ln}(N(t_2) - \mathrm{Ln}(N(t_1)]/(t_2 - t_1) = 0.0752 \text{ per day,per head.}
$$

---

[a] Any time can be chosen as the start, $t = 0$, so that the formula gets the simpler form: $N(t) = N(0)e^{rt}$.

Table F8 For explanation, see text.

Age group (8-day intervals)	0	1	2	3	4	5	7	9	11	13	15	20	24	28	48	64
Days (t)	0	8	16	24	32	40	56	72	88	104	120	160	192	224	384	512
Total (N)	1	10	20	26	29	54	332	765	2889	10,447	29,844	634,843	7,344,348	82,425,641	13,919,005,330,000	213,405,356,500,000,000
Fraction																
Eggs	0.0	90.4	51.1	25.2	14.4	47.7	55.2	33.0	55.2	45.1	44.7	45.4	45.9	47.3	47.0	47.0
Larvae	0.0	0.0	44.4	38.6	21.9	7.6	29.7	26.8	20.2	29.2	22.8	24.5	25.8	25.4	25.1	25.1
Protonymphs	0.0	0.0	0.0	33.3	33.3	11.5	7.3	22.6	8.3	14.4	14.9	14.4	14.0	13.1	13.3	13.3
Tritonymphs	0.0	0.0	0.0	0.0	28.6	17.5	1.2	12.1	6.6	5.2	9.6	8.2	7.1	6.7	7.0	7.1
Adults	100.0	9.6	4.5	2.9	1.8	15.6	6.6	5.6	9.7	6.1	8.0	7.5	7.2	7.4	7.5	7.5

Table F9 Extra information about life table B from Table 8.3a.

i Periods	Empirical Females	Empirical m_i	Empirical p_i	Hypothetical $p_i q$ ($q = 0.55$)	Hypothetical Females ($q = 0.55$)	Hypothetical m_i (m-Fraction $= 0.3$)
	65				65	
1	63	0	0.97	0.53	35	0
2	61	0	0.97	0.53	18	0
3	59	0	0.97	0.53	10	0
4	57	0	0.97	0.53	5	0
5	56	5.8571	0.98	0.54	3	1.7571
6	54	13.1667	0.96	0.53	1	3.9500
7	43	10.0698	0.8	0.44	1	3.0209
8	31	8.1935	0.72	0.4	0	2.4581
9	14	10.2500	0.45	0.25	0	3.0750
10	4	10.3750	0.29	0.16	0	3.1125
11	2	8.2500	0.5	0.28	0	2.4750
12	0	0	0	0	0	0

p_i = fraction escaping "intrinsic" mortality going from $i-1$ to i.
q = fraction escaping external causes of mortality (such as being sucked up by a vacuum cleaner) going from $i-1$ to i.
$p_i q$ = fraction surviving from $i-1$ to i.

F.4 Fabricated life tables

In order to clarify the procedure followed in Section 8.2. to construct some fabricated (hypothetical) life tables, an example will be presented in greater detail here (Table F9).

References

Arlian, L.G., Rapp, C.M., Ahmed, S.G., 1990. Development of *Dermatophagoides pteronyssinus*. J. Med. Entomol. 27, 1035–1040.

Birch, L.C., 1948. The intrinsic rate of natural increase of an insect population. J. Anim. Ecol. 17, 15–26.

Caswell, H., 1989. Matrix Population Models. Sinauer Associates, Inc., Publishers Sunderland, Massachusetts. ISBN 0-87893-094-9.

Colloff, M.D., 1987a. Mites from house dust in Glasgow. Med. Vet. Entomol. 1, 163–168.

Colloff, M.D., 1987b. Effects of temperature and relative humidity on development times and mortality of eggs from laboratory and wild populations of the European house dust mite *Dermatophagoides pteronyssinus* (Acari: Pyroglyphidae). Exp. Appl. Acarol. 3, 279–289.

Gamal-Eddin, F.M., Abou-Sinna, F.M., Tayel, S.E., Aboul-Atta, A.M., Seif, A.M., Gafaar, S.M., 1983. Duration of the development stages of house dust mites *Dermatophagoides farinae* and *D. pteronyssinus* under controlled temperatures and relative humidities to pave the way in front of the workers in the field of house-dust mite bronchia asthma. J. Egypt. Soc. Parasitol. 13, 319–334.

Matsumoto, K., Okamoto, M., Wada, Y., 1986. Effect of relative humidity on life cycle of the house dust mites, *Dermatophagoides farinae* and *D. pteronyssinus*. Jpn. J. San. Zool. 37, 79–90.

Appendix G

Dynamic modeling: STELLA

STELLA is a commercially available graphical simulation program. Two examples will be given to illustrate how it works.

Example G1

ARRIVE = MITES × 2

DT = 0.25

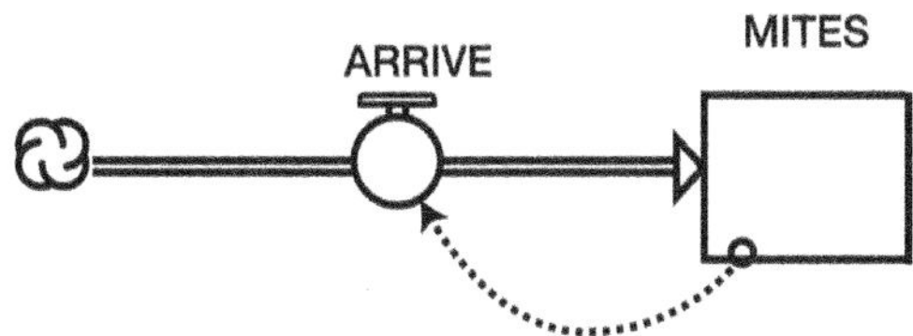

INITIAL MITES = 1

Time	Mites	Arrive	Incr. = Arrive × DT
0	1	2	0.5
0.25	1.5	3	0.75
0.5	2.25	4.5	1.125
0.75	3.375	6.75	1.6875
1	5.0625	10.125	2.53125

Example G2

$$N(t) = N(0)e^{rt}$$

$r = 0.52$ per head per week (Malthusian parameter)

$$N(t) = N(0)e^{0.52t}$$

$$N(t + \Delta t) = N(t)e^{0.52\Delta t}$$

$$N(t + \Delta t) - N(t) = N(t)e^{0.52\Delta t} - N(t) = N(t)\left(e^{0.52\Delta t - 1}\right)$$

- $N(t + \Delta t) - N(t) = \text{MITES} \times (\text{EXP}(0.52 \times \text{DT}) - 1)$
- $\text{ARRIVE} = (\text{MITES} \times (\text{EXP}(0.52 \times \text{DT}) - 1))/\text{DT}$

Appendix H

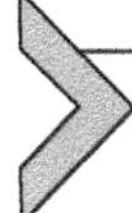

The binomial distribution, the poisson distribution, confidence intervals, and a dreamed sampling procedure

The model: a very large basket full of red and white beads thoroughly mixed. Samples, size n, are taken at random. The numbers of red beads in the samples will have a so-called binomial probability distribution.

If, for instance, you take samples of size $n = 10$, then the number of red beads in the samples is either 0, 1, 2, 3, 4, 5, 6, 7, 8, 9, or 10. It can be shown that the relative expected frequency, $fr(x)$, of each of these outcomes is as follows:

$$fr(x) = p^x \times (1 - p)^{n-x} \times n!/[x! \times (n - x)!]$$

whereas: the average $= \bar{x} = n \times p$ and the standard deviation $=$ s.d. $= [n \times p \times (1 - p)]^{0.5}$.

where:

$p =$ the fraction of red beads in the basket $(0 < p < 1)$

$n =$ the sample size (10 in this example)

$x =$ the number of red beads in the sample (either 0, 1, 2, 3, 4, 5, 6, 7, 8, 9, or 10)

For instance, when $p = 0.3$ the distribution is as presented in Table H1.

Table H1 Binomial distribution with $n = 10$ and $p = 0.3$.

x	$fr(x)$	$fr(\leq x)$
0	0.028	0.028
1	0.121	0.149
2	0.233	0.383
3	0.267	0.650
4	0.200	0.850
5	0.103	0.953
6	0.037	0.989
7	0.009	0.998
8	0.001	1.000
9	0.000	1.000
10	0.000	1.000

Note that the relative expected frequencies, $fr(x)$, add up to 1.

Example H1. You took a small dust sample from the total surface of a rug and recovered all the mites in it. The total number of live *Dermatophagoides pteronyssinus* was 12. Three of these were adults, the others were immature. So, based on this sample you estimate that the fraction of adults is 0.25 (25%). Following the directives in your statistical textbook, you find that the lower and upper limits of the 95% confidence interval (95% CI) of this estimate are 0.055 (5.5%) and 0.572 (57.2%), respectively.

Table H2 compares the distribution of x when p is exactly 0.055 (lower limit) with the distribution when p is exactly 0.572 (upper limit). This may clarify the meaning of the 95% confidence interval when you find three red beads in a sample of 12.

Table H2 Binomial distributions with $n = 12$ and $p = 0.055$ and $p = 0.572$, showing that, if you find 3 adults in a sample of 12 mites, the lower and upper limits of the 95% confidence interval of your estimate are 0.055 (5.5% adults) and 0.572 (57.2% adults), respectively.

		$P = 0.055$ (lower limit)		$P = 0.572$ (upper limit)	
x	x/n	$fr(x)$	$fr(x \geq 3)$	$fr(x)$	$fr(x \leq 3)$
0	0.000	0.507		0.000	
1	0.083	0.354		0.001	0.025
2	0.167	0.113		0.004	
3	**0.250**	**0.022**		**0.020**	
4	0.333	0.003		0.060	
5	0.417	0.000		0.128	
6	0.500	0.000		0.199	
7	0.583	0.000		0.228	
8	0.667	0.000	0.025	0.190	
9	0.750	0.000		0.113	
10	0.833	0.000		0.045	
11	0.917	0.000		0.011	
12	1.000	0.000		0.001	

Consider the model once more. Suppose that the proportion of red beads in the basket is very small. Only if you take a large sample you have a reasonable chance to find some red beads in it. But even when the proportion of red beads is nearly zero, you can still find some red beads in your sample if your sample is big enough. Mathematicians tell us that, in this situation, the binomial distribution of the numbers of red beads in your sample resembles a Poisson distribution. The usefulness of the Poisson distribution will be shown in the following example:

Example H2. Suppose you take a sample of dust (5 min vacuuming) from a rug ($16\,\text{m}^2$). The sample is placed for one or two days at 75% RH to allow viable mites to replenish their water reserves and gain a fully hydrated appearance. The larger and smaller dust particles are removed by sieving. All the retained fine dust is suspended in exactly 1 L of water (with a detergent). Under vigorous stirring, using a magnetic stirrer, 60 mL of the suspension is pipetted off and spread on black filter paper in a Buchner funnel under suction. This first subsample may serve as a pilot. Using a dissecting microscope you find 20 mites on the filter paper (red beads). In the suspension, all the space around these mites could have been filled with mites. But it was not. You can say that it was filled with an immense number of nonmites (white beads). Hence, your samples will be Poisson distributed.

The mean and the variance of a Poisson distribution are equal. (The conclusion in Appendix A2, that the mites in a certain carpet had a clustered distribution, is based on this property). Thus a Poisson distribution is defined by only one parameter, namely the mean. You found 20 mites in this subsample of 60 mL (95% CI: 13.0–30.9). That corresponds with a density of 21 mites/m^2 of carpet and a 95% CI ranging from 14 to 32 mites/m^2. You wish a narrower interval, which you would have if you captured more mites, for instance 100. Therefore you take two more subsamples from the suspension, each one measuring 120 mL. Now you find 34 and 41 mites, respectively. The three subsamples offer you the possibility to examine goodness of fit. The separate subsamples turn out to be in good agreement ($X^2 = 0.711$, d.f. $= 2$, $p \approx 0.7$). That is important because significant differences would have compromised the followed procedure of taking subsamples.

So you found a total of 95 mites (95% CI: 77.7–116.2) in the 300 mL which you took from 1 L suspension, containing the fine dust from 5 min vacuuming on $16\,\text{m}^2$ of carpet. This corresponds with a density of 20 mites/m^2 and a 95% CI ranging from 16 to 24 mites/m^2 of carpet, per 5 min vacuuming.

But there is more you can do or could have done.

You could have counted not only the mites but also the mite eggs on the black filter paper in the Buchner funnel. Then if you take at least two more dust samples from the rug (5 min vacuuming each), you can estimate the total number of mites and eggs separately through extrapolation (see Appendix A).

You could also have transferred the mites from the black filter paper onto a microscope slide in order to determine species and stage.

That will yield a picture of unprecedented quality of the structure of a mite population.

Index

Note: Page numbers followed by *f* indicate figures and *t* indicate tables.

CPI Antony Rowe
Eastbourne, UK
March 16, 2023